EDITED BY DR. SAROJ PACHAURI & DR. ASH PACHAURI

Unveiling

The Climate Change and Health Dynamic

Cover Design
Harun Ahmed

First edition

This book was professionally typeset on Reedsy.
Find out more at reedsy.com

Contents

Foreword

Today, talking about climate and health is an urgent necessity. Without a doubt a few decades ago, man still believed in the supremacy of human beings over the planet. However, recent experience with the COVID-19 pandemic has shown us that our species is capable of causing damage to nature in such a way that biodiversity becomes uncontrolled and disturbed. Our experience with zoonosis has made it clear that global actions that we undertake can impact, not only on biodiversity, but can also affect the resources that we dispose of indiscriminately.

This book represents a great step towards understanding the close relationship between the health of living beings and the environment that surrounds them. Diverse topics are addressed ranging from the conception of life and its relationship with nature to effects on human beings that are linked to the changing climate of the planet.

This book shows how for the *Yamaye Taino* indigenous peoples, their wellbeing is intrinsically linked to the health of their environment, which is part of a great network of life in which the health of the land, the spirits, plants, and animals directly impact human health. Wisdom, care, respect, and beliefs protect nature. Growing urbanization and industrialization generate contaminants which can harm public health. Increase in contaminants is linked to their regulation, which in developing countries is very lax. Their interrelationship with climate change causes important changes in the ecosystem resulting in the emergence of new microorganisms and modifications to climatic patterns. These interactions are poorly understood. Significant research efforts are needed to understand them. Increasing density of urban areas is impacting public health. Climate change generates heat stress, respiratory problems, and injuries, particularly in these areas that are characterized by the emission of high levels of greenhouse gases. Another vitally important topic discussed in this book is related to the challenges posed by using resources such as water and energy and their interconnections with health in large urban areas such as in Mexico City. Human ecocentric design represents a new vision for building environments that integrate human behavior and intervention with climate and health. Finally, a specific case of health and bone health is presented in the context of climate change and global warming which tells us how the climate is affecting bone health in the population.

In reading this book, we can get precise information which helps us to reflect on the role that each one of us plays in creating the problem. It also shows us ways in which we can contribute a little of ourselves to address the solution for this important problem.

Professor Norma Patricia Muñoz Sevilla
Instituto Politécnico Nacional (IPN)
Mexico

Preface

Climate change is causing a public health crisis. It has profound impacts on health. Climate change affects both physical and mental health directly and indirectly. For mitigating the adverse impacts of climate change on human health, it is important to understand the interlinkages between health and climate change.

Research-based evidence is needed to provide directions based on which programs and policies can be designed.

Recognizing it as an urgent need, the editors are publishing a series of books on health and climate change. This is the third of the series. Eminent scholars from developed and developing countries have contributed chapters to this book which address a wide range of issues related to health and climate change. These data are urgently needed to enhance our understanding of how climate change is impacting human health so that interventions to prevent its adverse health impacts can be designed.

Issues discussed by the authors of this volume include knowledge among the indigenous peoples who have had a long history of maintaining a balance between human health and the health of plants and animals which together form the web of life.

Climate change has resulted in the development of several contaminants which effect health. The far-reaching impact of contaminants on human health and the ecosystem are examined. The impacts of sand/dust/storms, urban heat islands, and flooding are discussed.

Human right to water and its connections with human health is examined along with the issue of commodification of water and its impact on the full enjoyment of the human right to water. The authors introduce the human eco-centric design concept and advocate for a built environment design that integrates human behavior, climate, and health.

The multifaceted impacts of climate change on human health are explored. Mitigation strategies including urban planning, technological advances, and targeted public health interventions to combat the problem are assessed.

Thus, the authors of this book delve into a range of problems related to climate change and human health and suggest innovate strategies for addressing the problem. A significant body of knowledge is provided by the authors to advance our efforts for mitigating the serious impacts of climate change on human health.

Acknowledgments

We are most grateful to Komal Mittal, Research Associate, Center for Human Progress, New Delhi, India and Youth Mentor, POP (Protect Our Planet) Movement, New York, USA for her incredible support as we went through multiple iterations of the chapters.

Komal worked untiringly to correct the changes that we made to improve the quality of each chapter. Her pleasant personality and cheerful demeanor made a great difference to our work on the book. She finalized the references of several chapters with the support of Drishya Pathak, Research Associate, Center for Human Progress, New Delhi, India and Youth Mentor, POP (Protect Our Planet) Movement, New York, USA and did the plagiarism checks for all the chapters.

Komal took considerable initiative with the authors of the book chapters which facilitated the process greatly.

1

Echoes in the Conch Shell: Taíno Warnings of a Changing Climate

Abstract

For the Yamaye Taino Peoples, our wellbeing is intrinsically linked to the health of their environment. They consider themselves to be a part of an enormous web of life in which the health of the lands, spirits, plants, and animals directly impacts their own health. The ancient way of life is based on the idea of balance. The increasing intensity of hurricanes, rising sea levels, and changes in rainfall patterns, disrupt this delicate balance. Climate change disrupts traditional food sources, medicinal plants, and natural cycles that we rely on for our wellbeing. These environmental changes have a direct impact on the health of the people. Food insecurity leads to malnutrition and related illnesses. Increased flooding exposes us to waterborne diseases. The mental and spiritual toll of witnessing the destruction of our ancestral lands cannot be overstated.

Traditional knowledge, threatened by climate change, may hold solutions. Passed down practices and knowledge of medicinal plants can be applied to adapt to the new environment. Climate change disproportionately harms those who have traditionally cared for the environment. It demands strong action from world leaders and calls for a global shift towards sustainable practices in order to preserve our planet's balance. We believe sharing our struggles can inspire others to reconnect with nature and fight to protect it. Only by working together can we overcome the problems caused by climate change and create a healthy future for everything on Earth.

Keywords: Climate change, indigenous knowledge, wellbeing, sustainability, traditional practices, environmental impact, food insecurity, cultural identity, medicinal plants, resilience.

Authors

Kasike 'Kalaan' Nibonrix Kaiman, Member of Council of Ancestral Indigenous Medicine of the Americas, Yamaye/Jamaica Council Of Indigenous Leaders (Y.C.O.I.L.), Caribbean Organization of Indigenous Peoples (C.O.I.P.), Caribbean Region Peace and Dignity Journeys Main Organizer [yukayekeguanija@gmail.com]

Kasikeiani Chieftainess Ronalda, Member of Yamaye/Jamaica Council of Indigenous Leaders (Y.C.O.I.L.), Caribbean Organization of Indigenous Peoples (C.O.I.P.), Caribbean Food For Climate Justice Research Group [revolutionoffoodinjamaica@gmail.com]

Introduction

"Indigenous stories, which are tied to lands, whether you are from the lands or not, help us to align with the cycles of these spaces and understand where our priorities are to be placed at whatever point in the cycle of life humanity is today."

- Kasike Kaiman

The authors explore the impact of climate change on the health and wellbeing of the *Yamaye Taino* and Caribbean Indigenous people from a cultural and traditional perspective. We weave together the scientific realities of a changing climate with Indigenous knowledge passed down through the generations, highlighting the inter-connectedness of humanity and nature.

The web of life

For the *Yamaye Taino* people, our health is intrinsically linked to the health of our environment. They see themselves as part of a vast web of life, where the wellbeing of plants, animals, and the spirits of the land directly affect their wellbeing. This concept of balance is central to their traditional way of life. They refer to a series of *Lokono* stories from Guyana, the distant ancestors of the *Yamaye Taino* people to highlight their ancient worldview.

Lokono creation story (as told by *Kasike Kalaan*)

In the beginning *Aiomun Kondi*, the Dweller in the Height, made the earth. But though the waves crashed upon the shores and the breezes moved over the land, no life was seen. Then *Aiomun Kondi* made the *Kumaka* known to *Taino* as the *ceiba* (silk-cotton) tree, to grow towards the clouds. From the *Kumaka* tree he scattered the branches. Those that fell in the water,

3

became fish and sea creatures. Those which remained in the air, became winged and feathered fowls. And those which fell on the ground, turned into beasts and reptiles, men, and women spreading from there to fill the whole earth.

At first, wild fruit were the only human food and water the only drink. No hunters roamed the forests destroying life. Beasts and birds lived alongside young children and none harmed or destroyed.

When the human race began to multiply, the Divine and ever-young *Wadili* saw that the daughters of man were fair and took them as wives. To the children that were born, he taught the three great arts, yet unknown- Fishing, Hunting and Tilling the Soil. These people were the *Lokono* that strangers called the *Arawaks*.

As time passed, the people forgot their stories. Forgot the grace of the creator and the beauty way. *Aiomun Kondi* saw deeds of blood and shame increase on the earth. He looked down and saw that the earth was now corrupt. In his fury he decided to remove mankind from the face of the earth.

One wise chief found favor and learned that fire would sweep the land. He was told to seek a reef of sand. This chief guided a small band to seek refuge. A deep and wide put was made and all the shrub and grass around it was burned. When they had completed building their refuge, sand clouds of smoke came rolling in. The trees on the edge of the clearing were engulfed in flames.

The men took refuge in the pit and the birds and beasts from the forest soon followed. When the flames expired and the smoke stopped, the small band went into the forest and behold the aftermath. What was once a beautiful forest was now a wasteland of ash and bones.

After many years, the people again forgot their stories. Forgot their history and forgot the strong lessons. And wickedness returned until the wrong seemed right and the right seemed wrong. Everyone was now selfish and did as they pleased.

Aiomun Kondi saw this. As the Great creator, he dreaded cleansing the land again so he gave a warning, "*Unless their evil ways ended, all would be destroyed by water*".

Marerewana was a good man, and he heeded the warning. He made a huge *canoa* (canoe). His neighbours mocked him. Yet he continued as guided. He worked from morning light till sunset to prepare the boat of refuge.

After many days, the canoe was finished. Yet, he feared that when the flood came, they would drift across the sea and he cried to *Aiomun Kondi* for answers. *Aiomun Kondi* told him to fasten his canoe to the great *Kumaka* tree and it would not drift from home. So, with long palm ropes, he fastened the boat to the *Kumaka* tree. During the flood, it safely weathered the storm.

After many days, the rain ceased and the waters receded. *Marerewana* saw that his boat rested by his home. After many years, the people again forgot their stories and now found themselves stricken with illness and drought. A young chief *Arawanili* lamented by a river over his inability to heal his people from the evil spirits that were causing illness, which was affecting the old and young alike.

Arawanili was visited by a water spirit *Orehu*. *Orehu* said "Tell me what causes such sorrow of your people? *Arawanili* replied "I am burdened by the torment of my people from the babe in arms to the old man. They are plagued by evil spirits with fever, pain, and sickness. If it were a human foe I could have a chance. But without a charm I can do nothing against these spirits and this drought". *Orehu* instructed *Arawanili* to take a branch and plant it on the hill. She told him that when the tree was ripe with fruit to bring two back to the river.

At the appointed time, the first two fruits that fell *Arawanili* brought back to *Orehu*. She showed him how to cut one of the fruits in half and to clean out the pulp from inside. The insides were good for conditions affecting the skin. She showed him how to bore holes in the top and bottom of the second fruit and to clean the inside. She gave him river stones to place inside of the fruit and also a stick to place inside the fruit as a handle which he could decorate with feathers.

This fruit was the *calabash*. *Arawanili* learned to make is the *maraca*. She told him to use the *maraca* with his chants and gifted him tobacco telling him to use the sacred plant with the *maraca* and his chants. The spirits causing illness would leave and the rains would return.

Arawanili was successful in healing his people and the land. It is said that when his time came for eternal rest, he was blessed to sit with *Aiomun Kondi* and look down with joy when the *Piamen* or as *Taino* would say *Behike* (medicine man) would use his charms against the spirits that cause pain and sickness. They say *Orehu* can still be seen today in quiet places on the river's bank.

This story is one of the many traditional stories still used today to provide several key lessons for our future generations:

1. **Consequences of imbalance**: These stories emphasize the importance of balance in nature. Balance is a central component in Indigenous philosophy which teaches us not to take more than we require and to live in harmony with our environment to ensure its longevity and sustainability. We see in these stories how, when humans forget their natural connection to the land, the natural world suffers. This reflects how climate change disrupts the delicate balance of our ecosystem and biodiversity. Floods and droughts are the direct consequences of a disrupted climate, Traditional food sources become threatened mirroring the impact of climate change on agriculture and food security.

2. **Loss of traditional knowledge:** We see how when we forget past experiences and lessons in our history, we become vulnerable to future disasters. Similarly, the world is seeing today that traditional knowledge about sustainable practices is invaluable in adapting to climate change.

3. **The cycles of consequences:** The stories repeat the patterns of humans forgetting their past, their place in the

7

world, their responsibility to the environment, and the consequences they suffer as a result. This reflects how ignoring climate change warnings has led to current problems and can lead to future problems.

4. **Importance of warnings and preparation:** *Marerewana*'s story emphasizes heeding warnings and taking proactive measures to prepare for potential disasters. This can be seen today as an early call for climate change.

5. **Nature's capacity for renewal:** Despite the devastation caused in the stories by fire, flood, and even drought, all hope is not lost. The story mentions the environment's ability to heal suggesting that even if climate change brings challenges, with effort we can restore balance and create a sustainable future.

6. **The value of traditional knowledge:** The stories culminate with the wisdom of the *Taino* peoples' ancestors the *Lokono* people. *Arawanili* learns from the water spirit *Orehu* how to use the *calabash* and tobacco for healing and create the first *Maraca*. The most important lesson for Indigenous peoples from these stories is the importance of preserving traditional knowledge and practices that can help us to adapt to environmental change.

For the indigenous people who hold these stories at the center of their cosmology and philosophy, they serve as powerful allegories for climate change. They remind us of the delicate balance between humans and nature, the consequences of greed, disregard for natural laws, and the importance of learning from the past and from those that are wisdom keepers of ancient knowledge, in order to create a sustainable future.

The increase in temperature, the greater number and frequency of extreme weather events, long droughts, more recurrent landslides and floods, increasing coastal erosion, and ocean acidification are increasingly everyday realities for Latin American and the Caribbean populations.

In Latin America and the Caribbean between 1998 and 2020, climate-related events and their impacts claimed more than 312,000 lives and affected more than 277 million people.

Although Latin America and the Caribbean only generate 10 percent of greenhouse gas emissions, they already suffer the worst effects of global warming. Cyclones, hurricanes, floods, droughts, rising sea levels, and loss of glaciers will generate more and more migratory movements and put the lives of millions of people in the region at risk both in cities and in the countryside. Climate change also affects the basic infrastructure, the supply of clean water, food production, and electricity generation. It puts the population's livelihoods and basic services at risk with losses and damage who economic value can exceed two percent of the annual gross domestic product (GDP).

A detailed examination of the impact of climate change on *Yamaye Taino* and the Caribbean Indigenous Peoples is provided below:

Climate change disrupts the balance

Increasing intensity of hurricanes, rising sea levels, and changes in rainfall patterns disrupt the delicate balance. This

9

disrupts our living conditions by displacing our livelihood and cultural practices of ceremonial rites. Traditional farming practices for our food sources are threatened by saltwater intrusion and unpredictable weather patterns. Continuous bouts of droughts and flooding places our medicinal plants in a constant state of struggle to survive in completely altered ecosystems. The changing climate disrupts the natural cycle we rely on for both physical and spiritual sustenance (1).

Health impacts: Specific impacts of climate change on the *Yamaye Guani Taino* people

Rising sea levels act like a silent thief, stealing away our beaches and slowly submerging our fishing villages. This displacement, not only forces families to relocate and rebuild their lives, but also damages and destroys sites that hold cultural significance for the people. These traditional spaces are where they connect with their ancestors and history. Their losses weaken the very fabric of our community.

The vibrant energy of our fishing villages, once bustling with activity, feel the impact of the receding shorelines. With less beach space, there are fewer visitors who come to purchase fresh fish and enjoy meals prepared at the traditional stalls. This decline in foot traffic translates to a significant loss of income for both fishers and those who rely on selling food to earn a living. The economic impact of environmental shifts creates a ripple effect throughout the community. The expansive beaches that were once a source of leisure and enjoyment for our people are slowly shrinking. The beauty of the white sand and the calming sound of the waves is being replaced by encroaching seawater. This loss extends beyond recreation. The beach plays a cultural role offering a place for islanders to relax, socialize, and connect with nature. Climate change disrupts this harmony and threatens our way of life.

Increased risk and fear: The rising sea level is more than just an inconvenience. It brings a constant sense of fear and danger. The close proximity of the water to the food stalls creates a very real possibility of flooding, disrupting the peaceful atmosphere, and jeopardizing the safety of our community members.

Disrupted farming cycles: The delicate balance of our traditional farming practices has been thrown into chaos by unpredictable rainfall patterns. Droughts parch the land, leaving crops withered and harvests meager. Conversely, intense floods soak the soil, leading to landslides and erosion that render farmlands unusable. These extremes disrupt the cycles that we have relied on for generations to cultivate food. This has created food insecurity and threatened the very foundation of our traditional way of life.

The heavy rains that come with erratic weather patterns are a double-edged sword. While they may provide some much-needed moisture, they also cause floods that saturate the soil, leading to landslides and land slippage. This damages the land itself making it difficult to cultivate crops. Periods of drought not only threaten crops but also necessitate government-imposed water restrictions. This lack of water availability makes it difficult to maintain proper hygiene and further jeopardizes the health of our organic crops.

Impact on medicinal plants

Climate change is forcing certain medicinal plants to relocate to cooler areas like the "duck plant" that has moved higher up the mountains. This shift disrupts our ability to access these vital plants in their traditional locations. These plants are not merely ingredients, they are part of our cultural heritage passed down through generations with a deep understanding of their specific locations and healing properties. Having to search for them in unfamiliar territories disrupts this knowledge and makes it difficult to practice our traditional medicine (2).

Some medicinal plants, like the *Chaney* root, are not cultivated but are harvested from the wild. The knowledge of where to find these plants and how to harvest them sustainably is traditionally held by the elders of our community. If these plants become scarce due to climate change, this invaluable knowledge may be lost along with them. This loss would not only impact our ability to heal ourselves using traditional methods, but also serves as a link to our cultural heritage.

Impact on cultural ceremonies

The extreme heat that accompanies climate change makes it difficult and sometimes dangerous for our elders and young people to participate in traditional ceremonies. These ceremonies, often held outdoors, have been the cornerstone of our culture, a way to connect with our ancestors and the natural world. However, intense heat forces us to take additional safety precautions, such as limiting ceremony duration or providing shade and hydration, which disrupts the traditional flow of the nature of these practices.

Ceremonies honoring the sun at midday for instance, now require additional safety measures to prevent heat stroke, a new concern for our people. This need to adapt disrupts the traditional way these ceremonies have been conducted for generations creating a new challenge for preserving our cultural heritage.

These specific examples highlight the far-reaching consequences of climate change on the *Yamaye Guani* people. It is not just a matter of environmental degradation, it is a threat to our cultural identity, our livelihood, and our very way of life. If we as Indigenous peoples can't eat what we grow, the loss will be more significant than we think.

Colonialism's displacement and harsh treatment significantly impacted our ancestors' health, leading to changes in food sources and traditional food processing methods. We have been seeing and feeling the effects of the vast change in how food is being processed. Jamaica's approach has undergone significant transformation. Gone are the days of planting and cooking traditional, healthy meals at home. Instead, we now eat at restaurants and most of the food on our plates has been outsourced. The increase of imported, processed foods for example: rice, flour, sugar, and canned products, has flooded stores and our homes. Our traditional methods of food preparation have become more distant from our hearts and minds as we venture farther from the kitchen. The prevalence of lifestyle diseases such as high blood pressure, diabetes, and cholesterol has escalated as statistics show.

This has laid the perfect foundation for a new rise of epistemic violence on our traditional foodways and farming by practices which we are still struggling to hold on to. Epistemic violence is telling us that highly processed foods such as canned food products are better than maize/corn that is home grown using natural traditional organic farming methods. Epistemic violence is silencing the voices of our ancestor's/family's time-honored traditions of food preparation that was geared towards strengthening families and building communities.

With the ever-growing impact of climate change continuing to ravage our already battered traditional systems, the medicines of the land that are integral to the nutrition of our physical bodies are now shifting away from us. This leads to the loss of our mental health stability.

All those affected by climate change, particularly Indigenous people, are forced to abandon their stable way of life. Our foundation is built upon the relationship with our ancestors throughout life. Going somewhere new and unfamiliar is not something that is readily accepted. And having to relocate and establish fresh relationships in a new place causes mental strain. In new spaces, there is no comfort because our ancestors are no longer surrounding us. Our children will not have the level of connection that they ought to and this contributes to a generational disconnect.

The mental health of our people is becoming increasingly destabilized with the constant worry of food insecurity and sovereignty which is leading to malnutrition and related illnesses. Increased flooding exposes us to water-borne diseases.

The mental and spiritual toll of witnessing the destruction of our ancestral lands cannot be overstated. This is leading to the major causes of our community's health break-down as the connection between our youth and elders are becoming more fragile. As youth will continue to seek new pathways for survival while leaving the elders and our traditional ceremonial and cultural way behind, it is resulting in a dissolute breakdown of our people.

Traditional knowledge offers solutions

Despite these challenges, our traditional knowledge and way of living offers invaluable insights for adapting to an ever-changing climate. Sustainable fishing and agricultural practices passed down through generations can be revisited and implemented. Our deep understanding of medicinal plants can inform the development of climate-resilient herbal remedies. This is fortified by our relationship with our land.

Cultivating resilience in a changing world

The *Yamaye Guani* people are not merely passive inheritors of this wisdom; they actively adapt their traditional practices to the challenges of climate change. Understanding the importance of *cassava*, a flood and drought resistant crop cultivated by our ancestors, the *Yamaye Guani* are reviving the cultivation of such traditional food sources. This really ensures food security in the face of unpredictable weather patterns demonstrating a practical application of knowledge passed down through the generations.

Ancestral gardens: Preserving knowledge for future generations

Collaboration with universities on projects to establish ancestral gardens in schools and communities reflects the *Yamaye Guani*'s commitment to safeguard their knowledge. These gardens serve as sanctuaries where traditional indigenous plants are cultivated fostering not only biodiversity, but also the dissemination of this knowledge to future generations. By ensuring the continuity of these practices, the *Yamaye Guani* weave a vital thread into the tapestry of ecological resilience.

Ceremonies: Rekindling the connection with the environment

The *Yamaye Guani* people maintain a deep connection to their environment through a vibrant calendar of traditional ceremonies held throughout the year. These ceremonies serve as more than cultural celebrations. They function as potent tools for generating environmental awareness. For instance, ceremonies honoring rivers in March, acknowledge the lifeblood of the land, while the *Yamaye Taino* New Year in May ushers in the rainy season, fostering a cyclical awareness of nature's patterns. Similarly, summer and winter solstices celebrated in June and December, respectively, reinforce the importance of living in harmony with the natural world's rhythms. These ceremonies act as bridges between the past and the present, reminding the community of their inter-connectedness with the environment.

Companion planting: A sustainable symphony

The *Yamaye Guani* people are reviving traditional practices like companion planting, a method where several crops are grown together. This polyculture approach fosters a symbiotic and biodiverse environment. By strategically planting nitrogen-fixing legumes alongside crops that deplete nitrogen from the soil or by interspersing taller plants with shorter ones to provide shade, companion planting creates a more resilient agricultural system. This not only optimizes land use but also allows crops to better withstand changing weather patterns, showcasing the adaptability inherent in traditional knowledge.

The *Yamaye Guani* people's traditional knowledge offers a powerful testament to the enduring wisdom of Indigenous cultures. By learning from their stories and adapting their practices, we gain valuable insights into navigating the challenges of climate change and ensuring a more sustainable future for all. As we face a future marked by environmental change, the *Yamaye Guani* people stand as a beacon of hope, reminding us that tradition and adaptation can go hand in hand, weaving a tapestry of resilience for generations to come.

Importance of ancestral connection: The broken web and its impact on mental wellbeing

For the *Yamaye Guani* people, the disruption of the balance with nature caused by climate change extends far beyond the physical realm. It severs a vital connection to their ancestors and the spiritual foundation of their wellbeing.

The disruption of the web

Traditionally, the *Yamaye Guani* have relied on the land and its bounty for not only physical sustenance, but also spiritual guidance. Medicinal plants were readily available. Whispers from the land revealed their location and the ceremonies connected them to a lineage of knowledge passed down through the generations. Climate change disrupts this web of inter-connectedness. When familiar plants become scarce, the ability to heal not just physical ailments but also maintain spiritual harmony diminishes. This loss of connection to the land that once guided them, creates a profound sense of unease and disorientation.

The *Yamaye Guani* way of life has been intricately woven into the natural cycles. Ceremonies held throughout the year mirrored the rhythm of the seasons reinforcing the ancestral connection to the environment. These ceremonies provided a space for not only celebration, but also for communion with the spirits and the wisdom of their forebears. Climate change disrupts these cycles making it difficult to observe traditional practices. This disconnect from the ancestral voices creates a sense of loss and disrupts the spiritual grounding that ceremonies provide.

Community breakdown and mental health challenges

The impact of climate change extends beyond the individual. As the environment becomes less hospitable, some people are forced to leave their ancestral lands severing family and community ties. This dispersal dismantles the intergenerational knowledge transfer that has sustained the *Yamaye Guani* people for generations. The loss of community support and cultural identity contributes to a rise in mental health challenges such as depression, anxiety, and substance abuse. These findings are mirrored globally with research highlighting the increased risk of mental health issues faced by Indigenous communities experiencing environmental degradation (3).

A call to protect the ancestral tapestry

Kasikeiani Ronalda emphasizes that our environment encompasses not just the physical world but also our mental, emotional, and spiritual wellbeing. It is within this space, interwoven with the family and the community, that cultural practices flourish. Protecting the environment is not just about safeguarding the land, it is about preserving the ancestral tapestry that sustains the *Yamaye Guani* people's spiritual health and cultural identity. By reconnecting with the land, sea, rivers, plants, animals, and mountains, we reconnect with the wisdom of our ancestors and weave a future where cultural practices and spiritual wellbeing can thrive once more.

The climate crisis is a global challenge, but its effects are felt most acutely by those who have lived in harmony with nature for generations. We urge world leaders to listen to our Indigenous people, to our elders and our knowledge keepers, and learn to work with us. Let us all take decisive actions to mitigate climate change. We call for a global shift towards traditional sustainable practices that respects the delicate balance of our planet.

Concluding comments

By sharing our experiences, we hope to inspire others to reconnect with our natural world and listen to its needs. Let us work together to fight for its protection. Only through collective, compassionate, empathetic, and sustainable actions can we weather the change and ensure a healthy future for all beings.

References

1.World Meteorological Organization. State of the Climate in Latin America and the Caribbean. World Meteorological Organization. 2023. https://wmo.int/publication-series/state-of-climate-latin-america-and-caribbean

2.Paul H. Williams. Jamaica's Taino chief speaks abroad on stories as medicine. The Gleaner. 2024 Apr 18. https://jamaica-gleaner.com/article/news/20240418/jamaicas-taino-chief-speaks-abroad-stories-medicine

3.Smye V, Browne AJ, Josewski V, Keith B & Mussell W. Social suffering: Indigenous peoples' experiences of accessing mental health and substance use services. International Journal of Environmental Research and Public Health. 2023 Feb 13;20(4):3288. https://www.ncbi.nlm.nih.gov/pmc/articles/PMC9958899/

2

Understanding the Interplay of Emerging Contaminants, Climate Change, and Public Health

Abstract

The dynamic relationship of emerging contaminants with climate change and its public health implications is complex. It requires careful examination and interpretation, because of the complexity of factors governing climate change and the ever-increasing list of emerging contaminants which are linked to urbanization and industrialization. Due to lax rules on control and compliance in developing countries, effective management of these emerging contaminants poses increasing management challenges. Besides mobilization of emerging contaminants, climate change also brings in new health risks due to changing ecosystems, inclement weather conditions and new microbes. These interlinkages are less understood. There are significant gaps in knowledge regarding the interlinkages between the emerging contaminants-climate change-health nexus. In this chapter, we have discussed the fate, transport, speciation, and toxicity of these emerging contaminants to address these knowledge gaps and have delved into advanced projections regarding future alterations in the global biogeochemical cycles of these contaminants. We have focused on their far-reaching consequences on both human health and ecosystem integrity. This analysis provides a comprehensive understanding of how these pollutants may interact with and influence various environmental processes and biological systems over time. We have recommended adaptive co-management models and comprehensive strategies based on scientific research and have suggested policy measures that would result in positive public health outcomes.

Keywords: Emerging contaminants, climate change, toxicity, polypropylene copolymer, bioaccumulation

Authors

Kriti Akansha, Research Scientist, Mu Gamma Consultants, Guru-gram, Haryana, India [kriti@mugammaconsultants.com]

Manisha Jain, Research Scientist, Mu Gamma Consultants, Guru-gram, Haryana, India [mjain@mugammaconsultants.com]

Introduction

The global economy drives the ongoing production and release of new chemical and biological agents that pose significant threats to global health and sustainability. Prior to the Industrial Revolution, naturally occurring pathogens like bacteria, fungi, and viruses were the primary contaminants of concern, endangering both ecosystem and human health. However, industrialization prompted major alterations in pollution patterns, adding new contaminants into the natural environment such as heavy metals, industrial chemicals, and particulate matter (1). According to the latest projections on climate change, global ecosystems will undergo significant and swift changes that will have unknown and unexpected effects on ecosystems and human health (2).

Heat waves can become more severe due to climate change because there is a greater likelihood of hot days and nights. Potential evapotranspiration will increase overall due to the rise in surface temperature caused by global warming (3)(4). Furthermore, a warmer climate's increased land evaporation might aggravate drought conditions, making the ecosystem more vulnerable to wildfires and lengthening their seasons (5). Because of the air's enhanced capacity to hold onto moisture, temperature changes in the atmosphere are associated with more significant precipitation events, like rain and snowstorms (6).

Climate change influences contaminant behavior, which can have significant consequences for ecosystems and the species within them. Heavy rainfall and flooding can overload storm tanks at wastewater treatment facilities, leading to the discharge of untreated wastewater into rivers and streams, which deteriorates freshwater quality. On the other hand, extended periods of drought can create water shortages, affecting local water usage. This can have significant impacts on industries such as food processing, where water scarcity may force a reduction in sanitation practices, potentially compromising food safety Flooding in areas with animal farms where antibiotics are overused may facilitate the spread of antibiotic-resistant bacteria into the surrounding environment (7).

Recent studies suggest a possible link between higher temperatures and the growing rates of antimicrobial resistance in human pathogens. Climate change is expected to prolong the intervals between major storms, leading to more intense and extended droughts, and to amplify the severity of storms when they do occur. This could result in the release and spread of pollutants, including emerging contaminants.

Future research should aim to more clearly delineate the connection between climate fluctuations and the emergence and reemergence of infectious diseases.

Emerging Contaminants and Climate Change Nexus

Changes in temperature and precipitation patterns

A changing climate can have an impact on the distribution and persistence of developing pollutants in the environment by changing patterns of precipitation and temperature. Certain pollutants may volatilize and degrade more quickly in warmer climates and their movement and destiny in soil and water systems may be affected by modifications in precipitation patterns (8,9). Changing temperatures and precipitation patterns enhance the likelihood of soil erosion and water contamination, among other consequences of climate change on soil and water systems.

Sea level rise and coastal contamination

One consequence of climate change is the rising sea levels, which can flood coastal areas and release pollutants previously held in sediments. As a result, there may be hazards to the health of people and marine species, as well as contamination of coastal ecosystems. Rising sea levels may cause persistent pollutants to resurface in coastal areas (10). Sea-level rise is causing coastal groundwater to rise in urban areas, contaminated soil, and altering the fate and transport of contaminants. Groundwater rise (GWR) is predicted to impact 18.1 million hectares of contaminated land in the US. Increase in groundwater elevation can mobilize soil contaminants, alter flow directions, and create new exposure pathways (11).

Climate change and ecosystems

Climate change is contributing to the loss of terrestrial biodiversity and is impacting ecosystem carbon storage, both directly and indirectly, through changes in land use. The negative ecological impacts of climate change are becoming more apparent and are likely to intensify over the coming decades, affecting precipitation variability, extreme weather events, and physiological stress on ecosystems (12). Climate change is affecting the world in several ways, including changing temperature and precipitation patterns, more frequent and severe extreme weather events, and altered physiological stress on ecosystems (13). These changes significantly impact the desert's biodiversity, including the loss of endemic species and changes in community composition (14).

Habitat loss and emerging contaminants

Climate change can exacerbate the impacts of major extinction drivers like habitat loss, contaminants, and invasive species, leading to conservation challenges and threats to biodiversity (15). Emerging contaminants, influenced by climate change can affect migratory patterns, feeding behavior, bioaccumulation, and biomagnification in food webs, posing challenges for monitoring and understanding their fate in the environment (16).

Climate change is leading to swift shifts in species distribution, phenology, and interactions with pests and diseases. It may also impose dispersal demands that exceed species' natural capacities, potentially disrupting ecosystems across various regions. Further research is necessary to explore species translocation, ecosystem engineering, and strategies for conserving rare species (15)

Mangroves ecosystems are vulnerable to the effects of both climate change and increasing contaminants. Sea level rise is the main consideration owing to their tidal nature. The effects of climate change encompass shifts in temperature, salinity, rainfall patterns, and rising greenhouse gas concentrations. Even temperature increases alone can lead to accelerated growth, reproduction, photosynthesis, and respiration, as well as alterations in community composition and biodiversity.

Mangrove forests, with their distinct physical, chemical, and ecological traits, may retain contaminants entering the ecosystem. The sediment characteristics of mangrove ecosystems influence contaminant retention, Persistent organic pollutants (POPs) can potentially adsorb to organic matter or become degraded through microbial communities (17). A case study on the mangrove ecosystem in the *Sundarbans*, Bangladesh, highlights the importance of mangroves in protecting coastal communities from the impacts of climate change. The Sundarbans is home to a diverse range of mangrove species and provide critical habitat for a wide range of marine species (18).

Fate and transport of emerging contaminants

The fate and movement of emerging contaminants' (ECs) in the environment are complex processes that are influenced by the physicochemical characteristics of the pollutants, the surrounding environment, and the interactions between the contaminants and the environmental media. Comprehending these processes is essential in evaluating the ecological and health hazards linked to ECs and in formulating effective control approaches.

Environmental pathways

Emerging contaminants can enter the environment in several ways:

- Landfill leachates:

Rainwater seeps into landfills and dissolves a complex mixture of industrial chemicals, medicines, and personal care products, which then move into nearby soils and groundwater (19).

• Industrial discharges:

In areas with lax regulatory control, manufacturing process effluents containing a variety of synthetic compounds, including solvents and heavy metals, are released into water bodies, contributing to the presence of emerging contaminants (20).

• Agricultural runoff:

Pesticide-treated fields and animal pastures where medications are administered contribute to the presence of emerging contaminants in water bodies through agricultural runoff during rainy events (21).

• Wastewater treatment effluent:

Despite treatment efforts, many wastewater treatment facilities are not completely successful in eliminating all pollutants, leading to the discharge of treated water with high concentrations of emerging contaminants into surface waters, which can adversely affect human health and aquatic ecosystems (19).

Surface water and groundwater

Pharmaceuticals and personal care products are frequently left in wastewater released into surface waterways because they are not completely removed during wastewater treatment. Research indicates that the effluent from wastewater

treatment plants often contains a variety of emerging contaminants, such as hormones, antibiotics, and other personal
care items. Compounds like polypropylene copolymer (PPCPs)
have the potential to alter biological processes and cause
bioaccumulation in aquatic organisms (21). PPCPs have the
potential to contaminate groundwater systems through soil
permeability, presenting long-term hazards. The persistence
and possible bioaccumulation of PPCPs in groundwater are
especially worrying due to their difficulty in remediation and
long-term exposure hazards. The transport behavior of PPCPs
in the subsurface is governed by regional geochemical and
hydrological conditions, including the properties of subsurface
media and the characteristics of the contaminants (22).

Soil and sediments

The fate and transport of ECs in soil and sediments are primarily
governed by their sorption to organic matter and interactions
with subsurface media, which can limit their mobility and lead
to the creation of localized pollution hotspots and long-term
persistence, posing risks to terrestrial ecosystems and the
food chain. ECs frequently stick to soil and sediment particles in terrestrial environments, limiting their mobility and
persistence. Pesticides and other hydrophobic contaminants
tend to attach tightly to organic materials in soil, decreasing
their mobility but enhancing their bioaccumulation and long-
term persistence (23). The sorption of hydrophobic ECs to
soil can create localized pollution hotspots, presenting long-

term exposure hazards due to high contaminant concentrations. As they move through the soil, nano-colloids and other emerging contaminants interact with soil particles and air-water interfaces with regional geochemical and hydrological conditions, that determine their transport behavior. Organic contaminants can interact with subsurface media through colloid filtration, affecting their mobility and fate. The potential mobility of heavy metals in soils is influenced by their chemical speciation and abundance in different geochemical phases, with higher mobility and bioavailability for metals in exchangeable and oxide-bound fractions. The long-term retention of hydrophobic ECs in soils can be hazardous to plants and soil organisms. They also have the potential to infiltrate the food chain. Understanding the sorption and mobility behavior of ECs in soils is crucial for predicting the risk of soil and groundwater contamination (24).

Health risks associated with climate change

The interaction between emerging contaminants and climate change poses difficult challenges for the environment and public health. Some examples of the many compounds include pharmaceuticals, personal care products, insecticides, and industrial chemicals that are classified as emerging contaminants because they have been found in the environment but are not yet subject to regulation. The occurrence, fate, and movement of these toxins may be affected by changes in environmental systems due to human activities, such as deforestation and burning fossil fuels.

Climate change impacts on vector-borne diseases

Climate change is altering the seasonality and geographic range of vector-borne diseases like dengue fever and malaria by affecting the distribution and abundance of disease vectors. Rising global temperatures and changing precipitation patterns are expected to modify the transmission of many vector-borne diseases and increase the likelihood of disease outbreaks in some areas. Vector-borne diseases contribute to over 17 percent of all infectious diseases, leading to more than 700,000 deaths each year and disproportionately affect the poorest populations. The lack of access to health services and clean water for almost half of the world's population is a major barrier to controlling the impacts of climate-sensitive vector-borne diseases (25)(26).

Food and water security

Climate change is affecting agricultural production through changes in temperature, precipitation patterns, and extreme weather events, leading to reduced crop yields and disruptions to the food supply. Climate change is also reducing freshwater availability, with declines in precipitation, glacier/snow melt, and groundwater depletion, especially in water-stressed regions. As freshwater resources become scarce, there may be increased reliance on alternative water sources like wastewater for irrigation and groundwater recharge. Usage of groundwater and untreated/partially treated wastewater can lead to introducing and accumulating emerging contaminants in the food chain. Emerging contaminants, such as pharmaceuticals, personal care products, and industrial chemicals can bioaccumulate and biomagnify through food webs, increasing human and ecosystem exposure. The presence of contaminants in water and food can compromise their quality and safety, further exacerbating food and water insecurity. Vulnerable populations, particularly in developing countries, are disproportionately affected by the combined impacts of climate change and contaminant exposure on food and water security (27).

Toxicity of emerging contaminants and potential impacts on human health and ecosystems

Emerging contaminants (ECs) include industrial chemicals, insecticides, personal care products, medicines, and their by-products that can potentially endanger the health of humans and ecosystems. These pollutants can have a variety of negative impacts, even at low concentrations, due to their special qualities and persistence. With the help of new research and findings, this section expounds on the detrimental effects of ECs on ecosystems and human health.

Impact on human health

The health of humans is seriously threatened by the presence of ECs in food, drinking water, and the environment (28)(29). Antibiotics, hormones, and other drugs are among the pharmaceuticals and personal care products that have been found in water systems across the globe. Even in tiny amounts, prolonged exposure to these substances can have a negative impact on health. Antibiotics pose a serious risk to public health, as they can lead to the emergence of germs that are resistant to them. Antibiotic resistance lowers the effective-ness of medications used to treat illnesses, resulting in more prolonged hospital stays, greater medical expenses, and higher death rates (30).

Hormone disruptors, which include some industrial chemicals and medications, can cause endocrine system disruption, which can result in immunological, developmental, and reproductive issues. Commonly found in plastics, chemicals like bisphenol A (BPA) and phthalates are known to mimic or suppress natural hormones, leading to disorders like decreased fertility, aberrant child development, and elevated cancer risks. Extended exposure to endocrine-disrupting chemicals (EDCs) can have deleterious effects on human health, especially for susceptible groups, including youngsters and pregnant women (31). Another worry is neurotoxicity, particularly when it comes to pollutants like some industrial chemicals and pesticides. Adults who are exposed to neurotoxic ECs are more likely to acquire neurodegenerative disorders, and children's cognitive development can be affected (Figure 1). For instance, children who are exposed to high levels of lead—even from environmental heritage sources—are more likely to experience aggressive behavior, concentration problems, and a lower intelligence quotient (32).

Many ECs, such as some industrial chemicals and pesticides, have been linked to cancerous consequences. Certain compounds, like polychlorinated biphenyls (PCBs) and certain per- and polyfluoroalkyl substances (PFAS), have been identified as likely human carcinogens and have been connected to several malignancies, including testicular, liver, and kidney cancers. Over time, a build-up of harmful effects from ongoing exposure to these carcinogens, even at low concentrations, can dramatically raise the risk of cancer (33).

Figure 1
Unveiling the risks of emerging contaminants: Understanding the growing threats to health and environment

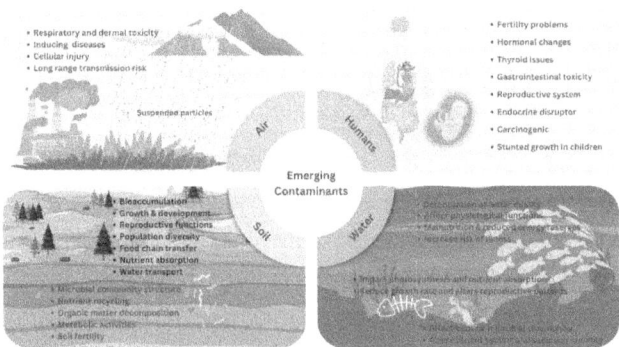

Source: *Implications of global climate change (1).*

Impact on ecosystems

Additionally, ECs severely harm ecosystems, having an impact on both terrestrial and aquatic settings. Since a significant number of ECs enter water bodies through industrial emissions, agricultural runoff, and wastewater discharges, aquatic ecosystems are especially vulnerable. Fish, amphibians, and invertebrates are among the aquatic creatures that may suffer from these pollutants. For example, drugs like hormones and antidepressants can affect how aquatic creatures behave, reproduce, and grow. Fish that are exposed to antidepressants such as fluoxetine, often marketed as Prozac, have been observed to exhibit altered behavior, such as a decreased propensity to avoid predators and an increased propensity to engage in risky activities (34)(35).

An additional serious issue with aquatic species is endocrine disruption. Fish population decreases and lower fertility can result from compounds like synthetic estrogens, which are present in birth control pills and other medications and can feminize male fish. Fish exposed to these endocrine disruptors have been shown to exhibit intersex abnormalities, in which male fish take on female traits, significantly reducing their chances of procreating (19).

ECs also have an impact on terrestrial ecosystems, especially when they contaminate soil. The persistence of pesticides and other hydrophobic pollutants in soils poses a long-term danger of exposure for soil species (36). The disruption of soil microbial communities by these pollutants can have an impact on soil health and nutrient cycling. Moreover, ECs can bioaccumulate in terrestrial food webs, endangering animals at higher trophic levels, such as mammals and birds (37).

Bioaccumulation and biomagnification are important EC-related concerns. Hydrophobic pollutants, which include several industrial chemicals and pesticides, tend to build up in the fatty tissues of living things. The concentrations of these pollutants can rise as they go up the food chain, exposing apex predators to greater amounts of exposure (38).

Management strategies and way forward

Analysis of gaps

- Current educational systems lack the necessary reforms to equip future generations with the skills and knowledge required to effectively tackle environmental and health challenges.
- Decision-makers often struggle to strike an appropriate balance between economic, social, and environmental considerations when developing policies for responsible chemical management.
- There is a significant gap in public awareness and comprehension regarding emerging contaminants, their origins, and potential consequences. Enhancing this knowledge is crucial for promoting responsible behavior and enabling informed decision-making that contributes to pollution reduction. It is essential to implement public education initiatives through various means such as awareness campaigns, interactive workshops, and community outreach programs. These efforts should focus on explaining the

scientific aspects of emerging contaminants, helping the public understand the associated environmental risks, and encouraging the adoption of environmentally conscious consumption habits.

· Addressing environmental pollution continues to be challenging due to the multifaceted nature of pollutants, limitations in available technologies, and obstacles in implementing comprehensive environmental policies. Ironically, chemical-based remediation techniques, which are often favoured, can sometimes cause more environmental harm than the pollution they aim to address. There is an urgent need to advance the fields of green chemistry, sustainable chemistry, and eco-friendly engineering to develop more effective and sustainable pollution prevention strategies for the future.

· It is crucial to acknowledge the interconnected nature of these environmental challenges to develop sustainable solutions that protect ecosystems, human health, and the well-being of future generations. This calls for integrated approaches that simultaneously address various aspects of global environmental change, supported by evidence-based policies and collaborative efforts across different sectors and disciplines.

Adaptive co-management models and comprehensive strategies

The relationship between emerging pollutants, climate change, and public health requires the implementation of comprehensive solutions and flexible co-management approaches. The creation of integrated monitoring systems that can continually detect the presence of toxins in different environmental matrices like water, soil, and air along with their effects on public health, is one essential strategy. To anticipate possible health concerns and enable prompt responses to new dangers, real-time data analytics and artificial intelligence should be used (39)

Effective co-management is based on community involvement and education. Strong community-based activities can be encouraged to raise public knowledge of the dangers posed by pollutants and climate change. Communities can be empowered, and a more robust response to environmental difficulties can be ensured by teaching local individuals how to monitor and report environmental dangers.

Innovation and cooperative research are also essential. New approaches to pollutant identification, mitigation, and remediation can be sparked by fostering collaborations between academic institutions, governmental organizations, and the commercial sector. A more thorough understanding of the relationship between environmental science, public health, and social sciences can be achieved through interdisciplinary study, which will enable more successful interventions (40).

The development of resilient infrastructure is another important tactic. Infrastructure that is resilient to the effects of climate change and capable of efficiently handling pollutants should receive funding. Resilience for the environment and public health can be enhanced by using green infrastructure solutions, such as built wetlands and green roofs, which have the dual benefits of filtering pollutants and reducing the effects of urban heat islands.

To manage and lessen the extent of emerging contaminants in the environment, strict controls are required in terms of policy and regulatory frameworks. It is necessary to develop and implement explicit standards for industries for the safe handling, disposal, and treatment of hazardous materials. To estimate potential health concerns under different climate scenarios, thorough health impact assessments (HIAs) must be carried out. Prioritizing public health concerns is ensured by including HIAs into urban planning and development projects (41).

Evidence-based policy recommendations for policy-makers and stakeholders

To successfully address the effects of emerging pollutants and climate change on public health, policy-makers and stakeholders at all levels must adopt evidence-based recommendations.

Data-driven decision-making

Policy-makers should use evidence-based data to guide their decisions, prioritizing activities that address the health implications of pollutants and climate change. Establishing centralized databases that aggregate research findings, monitoring data, and health statistics is essential for informed decision-making (42).

Interdisciplinary collaboration

It is important to encourage collaboration between scientists, medical professionals, and policy-makers to develop comprehensive strategies for controlling the pollutants-climate-health nexus. Frequent communication and information sharing between local, national, and international stakeholders will promote a coordinated and all-encompassing approach to addressing these issues.

Resource allocation

Governments and organizations need to commit enough money and resources to help with the monitoring, research, and mitigation of new pollutants and climate change. Allocating resources fairly to meet the needs of underserved and high-risk areas is crucial for advancing environmental justice and health equality.

Strong laws and activism

Comprehensive environmental and public health laws should be passed to properly address the issues brought on by new pollutants and climate change. Advocacy campaigns are necessary to increase public awareness and have an impact on legislative changes that support environment

Concluding comments

Emerging contaminants, climate change, and public health are interconnected issues that require a prompt and concerted response. Climate change, such as rising temperatures and changing precipitation patterns, can increase the toxicity, mobility, and bio-availability of pollutants, posing serious threats to ecosystems and public health. Climate change also increases the impacts of pollutants, causing more frequent and intense extreme weather events, spreading vector-borne diseases, and worsening respiratory ailments. The need for focused treatments disproportionately impacts vulnerable populations such as low-income communities, the elderly, and children.

To address these interconnected issues, an integrated approach that considers social, health, and environmental factors is essential. This includes adaptive co-management models, community involvement, cooperative research, resilient infrastructure, and strong monitoring systems. Evidence-based policies and regulations are needed to encourage sustainable behaviors and protect public health.

A coordinated call to action is necessary to promote public health and reduce the dangers associated with new pollutants and climate change. Policy-makers should prioritize strict laws, invest in green infrastructure, and ensure equitable resource allocation for vulnerable communities. Increased public participation and knowledge are also necessary, with education campaigns promoting a group response to environmental health risks.

In conclusion, tackling the relationship between new pollutants, climate change, and public health requires a multifaceted strategy that includes community involvement, policy development, and scientific study.

References

1.Wang F, Xiang L, Sze-Yin Leung K, Elsner M, Zhang Y, Guo Y, et al. Emerging contaminants: A One Health perspective. The Innovation. 2024 Jul 1;5(4):100612. https://doi.org/10.1016/j.xinn.2024.100612

2.Impacts, Adaptation and Vulnerability. Contribution of Working Group II to the Sixth Assessment Report of the Intergovernmental Panel on Climate Change. IPCC. 2022. https://www.ipcc.ch/2022/02/28/pr-wgii-ar6/.

3.Oelbermann M, Morgan S, Echarte L. Elevated carbon dioxide and temperature effects on soil properties from sole crops and intercrops. Soil Use Manag. 2022 Jan 20;38(1):435−47. https://doi.org/10.1111/sum.12752.

4.Peng Y, Xu H, Wang Z, Li L, Shang J, Li B, et al. Effects of intercropping and drought on soil aggregation and associated organic carbon and nitrogen. Soil Use Manag. 2023 Jan 26;39(1):316−28. https://doi.org/10.1111/sum.12866.

5.Abram NJ, Henley BJ, Sen Gupta A, Lippmann TJR, Clarke H, Dowdy AJ, et al. Connections of climate change and variability to large and extreme forest fires in southeast Australia. Commun Earth Environ. 2021 Jan 7;2(1):8. https://doi.org/10.1038/s43247-020-00065-8.

6.Trenberth K. Changes in precipitation with climate change. Clim Res. 2011 Mar 31;47(1):123–38. https://doi.org/10.3354/cr00953.

7.Bolan S, Padhye LP, Jasemizad T, Govarthanan M, Karmegam N, Wijesekara H, et al. Impacts of climate change on the fate of contaminants through extreme weather events. Vol. 909, Science of the Total Environment. Elsevier B.V.; 2024. https://doi.org/10.1016/j.scitotenv.2023.168388Get rights and content

8.Balbus JM, Boxall ABA, Fenske RA, McKone TE, Zeise L. Implications of global climate change for the assessment and management of human health risks of chemicals in the natural environment. Environ Toxicol Chem. 2013 Jan 18;32(1):62–78. https://doi.org/10.1002/etc.2046

9.Subramanian A, Nagarajan AM, Vinod S, Chakraborty S, Sivagami K, Theodore T, et al. Long-term impacts of climate change on coastal and transitional eco-systems in India: an overview of its current status, future projections, solutions, and policies. RSC Adv. 2023;13(18):12204–28. doi: 10.1039/d2ra07448f

10.Mimura N. Sea-level rise caused by climate change and its implications for society. Proceedings of the Japan Academy, Series B. 2013;89(7):281–301.DOI: 10.2183/pjab.89.281

11.Hill K, Hirschfeld D, Lindquist C, Cook F, Warner S. Rising Coastal Groundwater as a Result of Sea-Level Rise Will Influence Contaminated Coastal Sites and Underground Infrastructure. Earths Future. 2023 Sep 21;11(9). https://doi.org/10.1029/2023

EF003825

12.Malhi Y, Franklin J, Seddon N, Solan M, Turner MG, Field CB, et al. Climate change and ecosystems: threats, opportunities and solutions. Philosophical Transactions of the Royal Society B: Biological Sciences. 2020 Mar 16;375(1794):20190104. https://doi.org/10.1098/rstb.2019.0104

13.Ebi KL, Vanos J, Baldwin JW, Bell JE, Hondula DM, Errett NA, et al. Extreme Weather and Climate Change: Population Health and Health System Implications. Annu Rev Public Health. 2021 Apr 1;42(1):293–315.DOI: 10.1146/annurev-publhealth-012420-105026

14.Bombi P, Salvi D, Shuuya T, Vignoli L, Wassenaar T. Climate change effects on desert ecosystems: A case study on the keystone species of the Namib Desert Welwitschia mirabilis. Armas C, editor. PLoS One. 2021 Nov 8;16(11):e0259767. https://dx.plos.org/10.1371/journal.pone.0259767.https://doi.org/10.1371/journal.pone.0259767

15.Coristine LE, Kerr JT. Habitat loss, climate change, and emerging conservation challenges in Canada 1 This review is part of the virtual symposium "Flagship Species – Flagship Problems" that deals with ecology, biodiversity and management issues, and climate impacts on species at risk and of Canadian importance, including the polar bear (Ursus maritimus), Atlantic cod (Gadus morhua), Piping Plover (Charadrius melodus), and caribou (Rangifer tarandus). Can J Zool. 2011 May;89(5):435–51. https://doi.org/10.1139/z11-023

16.Bytingsvik J. Literature survey of emerging contaminants in the Barents Sea and climate change effects on contaminant fate. 2014. www.akvaplan.niva.no

17.Szafranski GT, Granek EF. Contamination in mangrove ecosystems: A synthesis of literature reviews across multiple contaminant categories. Mar Pollut Bull. 2023 Nov 1; 196:115595. DOI: 10.1016/j.marpolbul.2023.115595

18.Kumar Raha A, Zaman S, Sengupta K, Bikash Bhattacharya S, Raha S, Banerjee K, et al. Climate change–impact on the Sundarbans, a case study. papers.ssrn.com. 2013;107. DOI: 10.13140/2.1.2868.0328

19.Li X, Shen X, Jiang W, Xi Y, Li S. Comprehensive review of emerging contaminants: Detection technologies, environmental impact, and management strategies. Ecotoxicol Environ Saf. 2024 Jun; 278:116420. https://doi.org/10.1016/j.ecoenv.2024.116420

20.Wilson M, Aqeel Ashraf M. Study of fate and transport of emergent contaminants at waste water treatment plant. Environmental Contaminants Reviews. 2018 Jun 1;1(1):01–12. DOI: 10.26480/ecr.01.2018.01.12

21.Qiu L, Dong Z, Sun H, Li H, Chang C. Emerging Pollutants – Part I: Occurrence, Fate and Transport. Water Environment Research. 2016 Oct;88(10):1855–75.
DOI: 10.2175/106143016X14696400495811

22.Gani KM, Ali M, Dubey M, Kazmi AA, Kumari S, Bux F.

Transport of Emerging Contaminants from Agricultural Soil to Groundwater. In 2021. p. 261–81. DOI:10.1007/978-3-030-63249-6_10

23.Wen Y, Huang S, Qin Z, Chen Z, Shao Y. Priority screening on emerging contaminants in sediments of the Yangtze River, China. Environ Sci Eur. 2024 Feb 22;36(1):35. https://doi.org/10.1186/s12302-024-00855-3

24.Biel-Maeso M, Burke V, Greskowiak J, Massmann G, Lara-Martín PA, Corada-Fernández C. Mobility of contaminants of emerging concern in soil column experiments. Science of The Total Environment. 2021 Mar; 762:144102. https://doi.org/10.1016/j.scitotenv.2020.144102

25.World Health Organization. WHO factsheet vector-borne diseases. WHO 2014. https://extranet.who.int/kobe_centre/sites/default/files/pdf/vbdfactsheet.pdf

26.Marselle MR, Stadler J, Korn H, Irvine KN & Bonn A. Biodiversity and health in the face of climate change: challenges, opportunities and evidence gaps. In: Biodiversity and health in the face of climate change. Cham: Springer International Publishing. 2019:1–13. https://doi.org/10.1007/978-3-030-02318-8_1

27.Chakrabarty, Malana. Climate change and food security in India. Observer Research Foundation. 2016 Sep. https://www.orfonline.org/research/climate-change-and-food-security-in-india

28.Mani M, Pachaiappan R, Mahendra Gowda RV & Aroulmoji V. Environmental and Health Effects of Emerging Contaminants –A Critical Review. International Journal of Advanced Science and Engineering. 2023 Nov 25;10(2):3449–70. https://doi.org/ 10.29294/IJASE.10.2.2023.3449-3470

29.Lei M, Zhang L, Lei J, Zong L, Li J, Wu Z, et al. Overview of Emerging Contaminants and Associated Human Health Effects. Biomed Res Int. 2015;2015:1–12. https://doi.org/10.1155/2015/ 404796

30.Tarun Anumol. The Impact of Emerging Contaminants to the Environment and Human Health. Envirotech online. 2021 Dec 6;
  https://www.envirotech-online.com/article/water-wastew ater/9/agilent-technologies/the-impact-of-emerging-cont aminants-to-the-environment-and-human-health/3064

31.Schjenken JE, Green ES, Overduin TS, Mah CY, Russell DL, Robertson SA. Endocrine Disruptor Compounds—A Cause of Impaired Immune Tolerance Driving Inflammatory Disorders of Pregnancy? Front Endocrinol (Lausanne). 2021 Apr 12;12. https://doi.org/10.3389/fendo.2021.607539

32.Patel N, Khan ZA, Shahane S, Rai D, Chauhan D, Kant C, et al. Emerging pollutants in aquatic environment: Source, effect, and challenges in biomonitoring and bioremediation- A review. Pollution. 2020;6(1). https://doi.org/10.22059/POLL.2019.285 116.646

33.Yadav D, Rangabhashiyam S, Verma P, Singh P, Devi P,

Kumar P, et al. Environmental and health impacts of contaminants of emerging concerns: Recent treatment challenges and approaches. Chemosphere. 2021 Jun; 272:129492. https://doi.org/10.1016/j.chemosphere.2020.129492

34.Sultan MB, Anik AH & Rahman MdM. Emerging contaminants and their potential impacts on estuarine ecosystems: Are we aware of it? Mar Pollut Bull. 2024 Feb; 199:115982. https://doi.org/10.1016/j.marpolbul.2023.115982

35.Jiang JJ, Lee CL, Fang MD, Tu BW & Liang YJ. Impacts of Emerging Contaminants on Surrounding Aquatic Environment from a Youth Festival. Environ Sci Technol. 2015 Jan 20;49(2):792–9. https://doi.org/10.1021/es503944e

36.Nam K& Kim JY. Persistence and bioavailability of hydrophobic organic compounds in the environment. Geosciences Journal. 2002 Mar;6(1):13–21. https://doi.org/10.1007/BF02911331

37.Mishra RK, Mentha SS, Misra Y & Dwivedi N. Emerging pollutants of severe environmental concern in water and wastewater: A comprehensive review on current developments and future research. Water-Energy Nexus. 2023 Dec; 6:74–95. https://doi.org/10.1016/j.wen.2023.08.002

38.Gomes IB, Maillard JY, Simões LC & Simões M. Emerging contaminants affect the microbiome of water systems—strategies for their mitigation. NPJ Clean Water. 2020 Sep 18;3(1):39. https://doi.org/10.1038/s41545-020-00086-y

39.Maruya KA, Schlenk D, Anderson PD, Denslow ND, Drewes JE, Olivieri AW, et al. An adaptive, comprehensive monitoring strategy for chemicals of emerging concern (CECs) in California's aquatic ecosystems. Integr Environ Assess Manag. 2014 Jan 12;10(1):69–77. https://doi.org/10.1002/ieam.1483.

40.Plummer R, Baird J. Adaptive co-management for climate change adaptation: Considerations for the Barents Region. Sustainability. 2013 Feb 5;5(2):629–42. https://doi.org/10.3390/su5020629

41.Fabricius C, Currie B. Adaptive co-management. in: adaptive management of social-ecological systems. Dordrecht: Springer Netherlands; 2015:147–79. https://doi.org/10.1007/978-94-017-9682-8_9

42.Françoise M, Frambourt C, Goodwin P, Haggerty F, Jacques M, Lama ML, et al. Evidence based policy making during times of uncertainty through the lens of future policy makers: Four recommendations to harmonise and guide health policy making in the future. Archives of Public Health. 2022 Dec 18;80(1):140. https://doi.org/10.1186/s13690-022-00898-z

3

Climate Resilient Urban Planning: Pathways to Promoting Community Health and Wellbeing

Abstract

Climate change poses grave threats to global public health through direct impacts like heat stress, respiratory issues and injuries, and indirect effects stemming from social and economic disruptions. Urban areas are particularly vulnerable due to their high population density, concentrated infrastructure, and as major contributors to greenhouse gas emissions. The authors explore how urban planning strategies can enhance cities' resilience against the health impacts of three major climate change hazards— sand/dust storms, urban heat islands, and flooding. They highlight the need for planning interventions like developing green buildings, early warning systems, sustainable transportation, green infrastructure, and promoting urban densification and mixed-use neighborhoods to mitigate climate impacts, improve public health, reduce greenhouse gas emissions, and enhance overall resilience against climate-related disasters.

Keywords: Climate change, urban planning, sustainable development, public health, vulnerability

Authors

Kolade Victor Otokiti, Faculty of Spatial Sciences, University of Groningen, the Netherlands [otokitikolade@gmail.com]

Helen Abidemi Faturoti, Department of Urban and Regional Planning, Lagos State University, Nigeria [Helen.faturoti@lasu.edu.ng]

Introduction

Climate change poses an increasingly pressing threat to global public health. According to Banwell et al. (2018), the rising severity and frequency of climate change induced disasters causes both direct and indirect health risks. Direct health impacts include heat exhaustion, respiratory illnesses, injuries, and fatalities resulting from climate-related hazards like storms, droughts, heat waves, and floods. Indirect health effects stem from the social and economic consequences of disasters such as limited access to healthcare services, maternal deaths, malnutrition, food shortages, and long-term mental health issues. Furthermore, climate change is expected to impact biodiversity and the essential ecosystem services that are crucial for human health. Shifts in temperature and precipitation patterns may also influence the spread of disease carrying organisms such as those responsible for malaria and dengue fever, and water-borne diarrheal diseases. It is argued that even modest climatic alterations observed since the mid-1970s may already be accountable for an estimated 150,000 fatalities and approximately five million disability adjusted life years annually (1).

Cities play a crucial role in efforts to mitigate and adapt to climate change (2). Despite covering only about three percent of the earth's surface, cities consume approximately 80 percent of the world's energy, and are responsible for about 70 percent of global CO_2 emissions (3)(4). Urban areas face heightened vulnerability to climate change and natural disasters when compared to rural areas (5). With a significant number of the global population residing in cities and the presence of key infrastructure, urban areas are particularly susceptible to meteorological hazards and climate change impacts (6).

This 'driver and victim' relationship between cities and climate change impacts underscores the importance of cities in address-ing climate change impacts. Integrating climate considerations into urban planning strategies has been identified as a potent means of creating resilient cities that are better equipped to withstand and recover from climate-related disasters (7)(8). This study seeks to explore how urban planning strategies can enhance cities' resilience to the health impacts of climate change.

Climate change hazards and health impacts in urban areas

Sand and dust storms

Sand and dust storms (SDS) may be described as the elevation of small particles by solid and turbulent winds reducing ground level visibility to 1,000 meters or less. The process typically involves three phases: the entrainment or emission of surface material, its atmospheric transport, and its eventual deposition (9). SDS is considered a notable global environmental concern with far-reaching consequences including economic disruptions, closure of critical infrastructure like airports and schools, supply chain interruptions, crop destruction, and adverse health impacts on millions worldwide (10). They occur primarily in the world's drylands particularly in the "dust belt" spanning from the Sahara Desert across Central Asia to Northeast Asia. While most of these events transpire at low latitudes, approximately five percent of global desert dust emanates from high latitude sources including Greenland, Iceland, and the Patagonian Desert in Argentina (11).

The composition of sand and dust particles varies based on the source area. Mineralogically, they comprise mainly quartz, clay minerals, feldspar, plagioclase, calcite, iron oxides, and other components. At the same time, chemically they consist of silicon dioxide, aluminum oxide, iron oxides, calcium oxide, magnesium oxide, and potassium oxide, along with salts, organic matter, microorganisms, and pollutants from anthropogenic activities (12)(13).

Health impacts of sand and dust storms

Owing to their physical, chemical, and biological properties,

53

SDS impose both acute and chronic harm to humans. Inhaling respirable mineral dust, particularly that found in SDS, poses a substantial health hazard. Unlike many other pollutants, there is no safe threshold for exposure to these mixtures of organic and inorganic chemicals and substances as low levels can cause damage to health (14). For instance, exposure to organic dust can lead to the development of organic dust toxic syndrome (ODTS) in 30-40 percent of exposed individuals (15). Furthermore, chronic exposure to fine particulates may lead to premature death from cardiovascular and respiratory diseases such as lung cancer, acute lower respiratory infections, and pneumonia (14). Other respiratory disorders attributed to SDS include asthma, tracheitis, pneumonia, aspergillosis, allergic rhinitis, and nonindustrial silicosis commonly known as "desert lung" syndrome (16).

Furthermore, SDS may exacerbate existing health conditions and result in new ailments. Hashizume et al. (2020) argue that dust exposure can aggravate bronchitis, emphysema, cardiovascular disorders, and eye infections (17). Yitshak-Sade et al. (2015) note that dust storms can cause pediatric asthma emphasizing the vulnerability of children to these environmental hazards (18). Furthermore, SDS pollutants can result in skin irritation, meningococcal meningitis, valley fever, and diseases associated with toxic algal blooms. Reduced visibility during storms also increases the risk of mortality and injuries from highway accidents. Gross et al. (2018). Additional health risks include coughing, wheezing, lower respiratory tract infections, obstructive airway diseases, lung fibrosis, and interstitial lung disease.

Symptoms of SDS exposure often mimic flu-like symptoms such as general weakness, headache, chills, body aches, coughing, chest tightness, and shortness of breath. These symptoms can significantly impair individuals' daily functioning and quality of life. Exposure to dust endotoxins can also trigger an inflammatory response in the lungs leading to pulmonary dysfunction and immunological disturbances (15)(19).

While the health consequences of SDS are global, populations in arid and near arid zones are particularly vulnerable. Long-range atmospheric transport can carry dust particles far from their source regions exposing distant populations to health risks. For example, Asian dust has been shown to contribute to aerosol loadings in western North America and African dust transported to the Caribbean and Florida has led to violations of the air quality standards of the United States (18).

Drivers of storm dust

Human activities represent an additional source of dust storms along the margins of deserts in semi-arid regions disrupting previously stable ecosystems. The most significant threat to these natural systems stems from human interference in or near areas highly prone to dust emissions. In regions like west and north-west India, locally referred to as *andhi*, the south-west coast of the Persian Gulf and the west coast of the Red Sea known as *haboob* as well as in the Northern Sahara, Sahel, and Mali termed *harmattan*, the frequency and intensity of dust storms has surged over the past decade. This surge correlates with the rising human population in these areas expanding desertification across larger land masses. Unsustainable agricultural practices, environmental degradation, and the accelerating pace of industrialization further drove this expansion. Additionally, human-induced alterations to hydrological systems resulting in the drying of wetlands or ephemeral water bodies further elevates the risk of severe SDS (16).

Urban heat islands

The world is experiencing increasing urbanization with urban areas expanding from 0.23 percent in 1992 to 0.53 percent in 2013 (20). By 2050, approximately two-thirds of the world's population is projected to reside in urban settings. This urbanization trend has given rise to a phenomenon known as the urban heat island (UHI), which is characterized by elevated temperatures in urban areas compared to their rural surroundings (21)(22). According to Choi et al. (2014), approximately 65 percent of global urban residents experience the UHI effect significantly impacting local economies and potentially leading to substantial GDP losses by 2100 compounded by future climate change (23).

Health impacts of urban heat islands

The UHI effect poses diverse challenges to urban microclimates affecting human health, labour productivity, and biodiversity (24). Its impact on human health is multifaceted and is driven by extreme temperatures, heat waves, and air pollution (25). Severe heat events coupled with reduced nocturnal cooling and diminished vegetation may exacerbate morbidity and mortality, allergic respiratory diseases, and worsen existing medical conditions especially among vulnerable populations (26)(27). Furthermore, UHI contributes to increased energy consumption exacerbating air pollution and greenhouse gas emissions which are linked to adverse cardiovascular and respiratory effects (28)(29).

Extreme temperatures resulting from UHI can affect multiple organ systems beyond the cardiorespiratory system causing intravascular dehydration, renal failure, and cognitive dys-function (30). Additionally, UHI causes discomfort leading to symptoms like lack of concentration, exhaustion, dehydration, and circulatory disorders (31). The burden of outdoor air pollu-tion, exacerbated by UHI, disproportionately affects low- and middle-income countries contributing to millions of deaths annually from various respiratory and cardiovascular diseases (32).

Drivers of urban heat islands

Cities are pivotal in driving global environmental transfor-mation primarily due to the widespread expansion of imper-meable surfaces worldwide. Coupled with increasing global temperature, the rapid growth of urban populations has led to significant alterations in urban and rural landscapes. This transformation is characterized by extensive changes in land use and land cover (LULC), particularly the replacement of natural vegetation and green spaces with impermeable surfaces such as concrete, asphalt, parking lots, rooftops, and building walls (31).

These LULC changes significantly impact surface temperatures in urban and surrounding areas by altering near surface energy budgets, increasing evapotranspiration, and creating surfaces that absorb solar energy, thus generating elevated surface heat levels. The rapid urbanization process, marked by expanding impermeable surfaces, contributes to the intensification of the urban heat island (UHI) effect. These biophysical materials exhibit high heat reemission capacities further exacerbating warming trends in the surrounding urban environment (33).

Flooding

Floods are among the most devastating natural phenomena impacting millions globally. Over the past three decades, floods have emerged as the most catastrophic natural disaster affecting an average of approximately 80 million individuals annually which accounts for half of the total population affected by any natural disaster worldwide. Floods pose significant threats to property, agriculture, and public utilities leading to billions of dollars in damage each year and the devastating loss of human and animal lives (34). Unfortunately, the intensity and frequency of flood occurrences are expected to increase in the coming decades (35).

Health impact of flooding

Flooding imposes manifold health implications on humans (36). Physically, flooding can lead to injuries, outbreaks of infectious diseases, malnutrition, decreased birth rates, and exacerbation of chronic illnesses (36). The psychological impact of flooding includes post-traumatic stress disorder (PTSD), anxiety, depression, distress, insomnia, nightmares, and suicidal ideation.

Flooding threatens the accessibility of vital health services, equipment, clean water, and food exacerbating existing health conditions, particularly psychological illnesses and chronic diseases (37). Beyond immediate health consequences like trauma, drowning, and poisoning, the aftermath of flooding carries long-term health risks arising from exposure to flood-waters, injuries, water contamination, food shortages, and the stress of cleanup and reconstruction efforts (36). Such long-term physical health impacts include asthma, gastrointestinal ailments, and increased rates of acute myocardial infarction (36)(38). Additionally, flooding is associated with reduced health-related quality of life, disability adjusted life years, acute myocardial infarction, and malnutrition (36).

Drivers of floods

Flooding is a complex phenomenon with various contributing factors. It is often initiated by rivers overflowing beyond their banks. This overflow can result from excessive precipitation and limited channel capacity. Additionally, obstructions within river beds can impede the natural water flow exacerbating the risk of flooding. Furthermore, flooding can occur at the confluence of streams where elevated river levels cause the water to back up into tributaries and surrounding areas (34).

The causes of flooding extend beyond river overflow to include a multitude of natural and human-induced factors. High rainfall, tidal extremes, tsunamis, cyclones, hurricanes, and rising sea levels are natural phenomena that can lead to flooding (39).

Additionally, changes in rainfall, temperature patterns, coastal inundation, and erosion contribute to flood risk (36).

Human activities also significantly contribute to exacerbating flooding. Poor physical planning, flaws in drainage networks, rapid urbanization, and informal housing development practices contribute to increased vulnerability to floods (40). The rapid expansion of cities and inadequate infrastructure and waste management practices may further compound the problem by disrupting natural hydrological processes (41)(29).

Urban planning strategies for addressing the impact of climate change on health

Cities offer critical opportunities to address the severe impacts of climate change. The building sector accounts for approximately 40 percent of total energy consumption and contributes to about one-third of greenhouse gas emissions in many countries (42). Through planning laws and regulations, planners can foster the construction of environmentally-friendly buildings that require less energy for air conditioning and are fitted with energy and water efficient appliances. The potential of green buildings to address the impacts of climate change and produce health co-benefits has been well documented in the literature (43)(44). Furthermore, early warning systems and monitoring combined with vulnerability and risk impact assessment and mapping can potentially reduce morbidity and prevent mortality (45).

Additionally, the development of new transportation (or upgrading of existing) infrastructure to promote low carbon modes of travel presents another opportunity for planners to address the impact of climate change as transportation contributes to 23–30 percent of global energy related greenhouse gas emissions (46). This may be complemented by sustainable urban development that encourages walkable neighborhoods, mixed-use development, densification, and reduced energy use. Sustainable urban development not only contributes to climate change mitigation but also improves urban livability and health (47)(48).

Green infrastructure, such as green roofs, urban forests, rain gardens, and bioswales play a crucial role in improving microclimates and mitigating the impacts of climate change (49)(50)(51)(52). Nieuwenhuijsen (2021) posits that green infrastructure offers several health benefits by reducing air pollution and heat exposure and improving mental health, reducing stress, improving the immune response, and active living.

Concluding comments

In light of the increasing frequency and severity of climate change-induced disasters, the direct and indirect health impacts on urban populations are becoming a growing concern. Addressing the health impacts of climate change in urban areas requires a comprehensive and integrated approach to urban planning. By adopting and scaling up planning strategies, cities can enhance their resilience, protect public health, and contribute to global climate change mitigation efforts. Policymakers and urban planners must prioritize strategies to create sustainable, healthy, and resilient urban environments. Future research should focus on evaluating the long-term outcomes of these interventions and explore innovative solutions to emerging climate-related health challenges.

References

1.Patz JA & Olson SH. Climate change and health: global to local influences on disease risk. Annals of Tropical Medicine & Parasitology. 2013 Jul 18, 100: 5-6; 535-549. https://www.tan dfonline.com/doi/abs/10.1179/136485906X97426

2.Heidrich O, Reckien D, Olazabal M, Foley A, Salvia M, de Gregorio Hurtado S, et al. National climate policies across Europe and their impacts on cities strategies. Journal of Environmental Management. 2016 Mar 01; 168:36−45. https://www.sciencedi rect.com/science/article/pii/S0301479715303972

3.World Bank. Cities and climate change: An urgent agenda. World Bank. 2011 Sep 07. https://documents.worldbank.org/e n/publication/documents-reports/documentdetail/19483146 8325262572/Cities-and-climate-change-an-urgent-agenda

4.Gouldson A, Colenbrander S, Sudmant A, Papargyropoulou E, Kerr N, McAnulla F, et al. Cities and climate change mitigation: Economic opportunities and governance challenges in Asia. Cities: Elsevier. 2016 May 01; 54:11−19. https://www.sciencedi rect.com/science/article/pii/S0264275115001638

5.Loorbach D, Shiroyama H, Wittmayer JM, Fujino J & Mizuguchi S. Theory and practice of urban sustainability transitions. Springer Book Series. https://www.springer.com/

series/13408

6.Masson W. Lemonsu A, Hidalgo J & Voogt J. Urban climates and climate change. Annual Reviews. 2020 Aug 14. https://www.annualreviews.org/content/journals/10.1146/annurev-environ-012320-083623

7.Hendricks MD & Zandt S V. Unequal protection revisited: Planning for environmental justice, hazard vulnerability, and critical infrastructure in communities of color. Environmental Justice. 2021 Apr 16; 14:2. https://www.liebertpub.com/doi/abs/10.1089/env.2020.0054

8.Abubakar IR & Dano UL. Sustainable urban planning strategies for mitigating climate change in Saudi Arabia. Environment, Development and Sustainability. 2020 Aug 01; 22(6):5129–52. https://doi.org/10.1007/s10668-019-00417-1

9.World Meteorological Organization. WMO airborne dust bulletin. World Meteorological Organization. 2020 May; 4. https://www.mgm.gov.tr/FTPDATA/arastirma/toz/sdswa/WMO_Airborne_Dust_Bulletin_No4_en.pdf

10.Cao H, Amiraslani F, Liu J & Zhou N. Identification of dust storm source areas in West Asia using multiple environmental datasets. Science of the Total Environment. 2015 Jan 01; 502:224–35. https://www.sciencedirect.com/science/article/pii/S0048969714013357

11.Wiley. High-latitude dust in the Earth system - Bullard - 2016. Reviews of Geophysics - Wiley Online Library. 2016. https://ag

upubs.onlinelibrary.wiley.com/doi/10.1002/2016RG000518

12.Middleton N & Kang U. Sand and dust storms: Impact mitigation. Multidisciplinary Digital Publishing Institute. 2017 Jun 17. https://www.mdpi.com/2071-1050/9/6/1053

13.Prokofieva T, Kiryushin AV, Shishkov VA & Ivannikov FA. The importance of dust material in urban soil formation: the experience on study of two young Technosols on dust depositions. Journal of Soils and Sediments. 2024 Oct 22. https://www.researchgate.net/publication/307951792_The_importance_of_dust_material_in_urban_soil_formation_the_experience_on_study_of_two_young_Technosols_on_dust_depositions

14.Naddafi K, Atafar Z, et. al. Health effects of airborne particulate matters (pm10) during dust storm and non-dust storm conditions IN TEHRAN. Journal of Air Pollution and Health. 2016: 1(4). https://japh.tums.ac.ir/index.php/japh/article/view/74

15.Achudume A & Oladipo B. Effects of dust storm on health in the Nigerian environment. Biology and Medicine. 2009 Jan 1;1. https://www.walshmedicalmedia.com/open-access/effects-of-dust-storm-on-health-in-the-nigerian-environment-0974-8369-1-039.pdf

16.Shepherd G, Terradellas E, Baklanov A, Kang U, Sprigg W, Nickovic S, et al. Global assessment of sand and dust storms. United Nations Environment Programme. 2016. https://repositorio.aemet.es/handle/20.500.11765/4495

17.Hashizume M, Kim Y, Ng CFS, et. al. Health effects of asian dust: A systematic review and meta-analysis. Environmental Health Perspectives. 2020 Jun 26; 128(6). https://ehp.niehs.ni h.gov/doi/full/10.1289/EHP5312

18.Yitshak-Sade M, Novack V, Katra I, Gorodischer R, Tal A & Novack L. Non-anthropogenic dust exposure and asthma medication purchase in children. The European Respiratory Journal. 2015 Mar;45(3):652–60. https://pubmed.ncbi.nlm.ni h.gov/25323244/

19.Ortiz-Martínez MG, Rodríguez-Cotto RI, Ortiz-Rivera MA, Pluguez-Turull CW & Jiménez-Vélez BD. Linking endotoxins, African dust PM10 and asthma in an urban and rural environment of Puerto Rico. Mediators of Inflammation. 2015(1):784212. https://onlinelibrary.wiley.com/doi/abs/ 10.1155/2015/784212

20.Dewan A, Kiselev G, Botje D, Mahmud GI, Bhuian Md & Hassan QK. Surface urban heat island intensity in five major cities of Bangladesh: Patterns, drivers and trends. Sustain Cities and Society. 2021 Aug 01; 71:102926. https://www.scien cedirect.com/science/article/pii/S2210670721002122

21.Yue W, Liu X, Zhou Y & Liu Y. Impacts of urban configuration on urban heat island: An empirical study in China mega-cities. Science of the Total Environment. 2019 Jun 25; 671:1036–46. https://www.sciencedirect.com/science/article/pii/S004896 9719314329

22.Semudara OM. Onibaba PO, Famewo AS & Otokiti KV.Impact

of urban expansion on urban heat: A case study of greater London. SpringerLink. 2024 May 22. https://link.springer.com/chapter/10.1007/978-3-031-57456-6_9

23.Choi Y-Y, Suh M-S, Park K-H. Assessment of surface urban heat islands over three megacities in East Asia using land surface temperature data retrieved from COMS. 2014 Jun 20. https://www.mdpi.com/2072-4292/6/6/5852

24.Tian P, Li J, Cao L, Pu R, Wang Z, Zhang H, et al. Assessing spatiotemporal characteristics of urban heat islands from the perspective of an urban expansion and green infrastructure. Sustainable Cities and Society. 2021 Nov 01; 74:103208. https://www.sciencedirect.com/science/article/pii/S2210670721004868

25.Singh N, Singh S & Mall RK. Chapter 17 - Urban ecology and human health: implications of urban heat island, air pollution and climate change nexus. In: Verma P, Singh P, Singh R, Raghubanshi AS, editors. Urban Ecology. Elsevier. 2020: 317–34. https://www.sciencedirect.com/science/article/pii/B9780128207307000173

26.Buchin O, Hoelscher MT, Meier F, Nehls T & Ziegler F. Evaluation of the health-risk reduction potential of countermeasures to urban heat islands. Energy and Buildings. 2016 Feb 15; 114:27–37. https://www.sciencedirect.com/science/article/pii/S0378778815300657

27.Kouis P, Kakkoura M, Ziogas K, Paschalidou AK & Papatheodorou SI. The effect of ambient air temperature on cardiovascular and respiratory mortality in Thessaloniki, Greece. Science of the Total Environment. 2019 Jan 10; 647:1351–8. https://www.sciencedirect.com/science/article/pii/S004896 9718330705

28.Giridharan R & Emmanuel R. The impact of urban compactness, comfort strategies and energy consumption on tropical urban heat island intensity: A review. Sustainable Cities and Society. 2018 Jul 01; 40:677–87. https://www.sciencedirect.co m/science/article/pii/S2210670717313446

29.Otokiti KV, Akinola O & Adenji KN. Geospatial mapping of flood risk in the coastal megacity of Nigeria. Research Gate 2020 Feb. https://www.researchgate.net/publication/339302 054_Geospatial_Mapping_of_Flood_Risk_in_the_Coasta l_Megacity_of_Nigeria

30.Walter EJ & Carraretto M. The neurological and cognitive consequences of hyperthermia. Critical Care. 2016 Jul 14. https://link.springer.com/article/10.1186/s13054-016-1376- 4

31.Ward K, Lauf S, Kleinschmit B & Endlicher W. Heat waves and urban heat islands in Europe: A review of relevant drivers. Science of the Total Environment. 2016 Nov. https://www.scie ncedirect.com/science/article/abs/pii/S0048969716312931

32.Wong LP, Alias H, Aghamohammadi N, Aghazadeh S & Nik Sulaiman NM. Urban heat island experience, control measures and health impact: A survey among working community in the city of Kuala Lumpur. Sustainable Cities and Society. 2017

Nov 01; 35:660–8. https://www.sciencedirect.com/science/article/pii/S2210670717307904

33.Balew A & Semaw F. Impacts of land-use and land-cover changes on surface urban heat islands in Addis Ababa city and its surrounding. Environment Development and Sustainability. 2022 Jan. https://www.researchgate.net/publication/352696 850_Impacts_of_land-use_and_land-cover_changes_on _surface_urban_heat_islands_in_Addis_Ababa_city_and _its_surrounding

34. Orimoloye IR, Adefisan EA & Abdulkareem S. Application of geo-spatial technology in identifying areas vulnerable to flooding in Ibadan metropolis. Virchows Archiv. 2015 Jan. https://www.researchgate.net/publication/226569363_Appl ication_of_Geo-Spatial_Technology_in_Identifying_Area s_Vulnerable_to_Flooding_in_Ibadan_Metropolis

35.Bloschl G, Gaal L, Kiss A, et. al. Increasing river floods: Fiction or reality? WIREs Wiley Online Library. 2015 Mar 11. https://wires.onlinelibrary.wiley.com/doi/10.1002/wat2.1079

36.Zhong S, Yang L, Toloo S, Wang Z, Tong S, Sun X, et al. The long-term physical and psychological health impacts of flooding: A systematic mapping. Science of the Total Environment. 2018 Jun 01; 626:165–94. https://www.scien cedirect.com/science/article/pii/S0048969718300494

37.Bland SH, O'Leary ES, Farinaro E, Jossa F & Trevisan M. Long-term psychological effects of natural disasters. Psychosomatic Medicine. 1996 Jan-Feb;58(1):18–24. https://pubmed.ncbi.nl

m.nih.gov/8677284/

38.Jiao Z, Kakoulides SV, Moscona J, Whittier J, Srivastav S, Delafontaine P, et al. Effect of hurricane katrina on incidence of acute myocardial infarction in New Orleans three years after the storm. The American Journal of Cardiology. 2012 Feb 15;109(4):502−5. https://www.sciencedirect.com/science/article/pii/S0002914911030384

39.Adeniran I & Otokiti KV. Characterization of climate change manifestation in Nigeria's coastal communities. 2019 Jul. Research Gate. https://www.researchgate.net/publication/334576150_Characterization_of_climate_change_manifestation_in_Nigeria's_coastal_communities

40.Amoako C & Boamah EF. The three-dimensional causes of flooding in Accra, Ghana. International Journal of Urban Sustainable Development. 2015 Jan 02. https://www.tandfonline.com/doi/abs/10.1080/19463138.2014.984720

41.Adeniran I, Otokiti K & Durojaye P. Climate change impacts in a rapidly growing urban region -A case study of Ikeja, Lagos, Nigeria. 2020 Mar 1; 6:13−23. https://www.researchgate.net/publication/339642976_Climate_Change_Impacts_in_a_Rapidly_Growing_Urban_Region_-A_Case_Study_of_Ikeja_Lagos_Nigeria

42.UN Habitat. Global report on human settlements 2011. Cities and climate change. UN Habitat. 2011. https://unhabitat.org/global-report-on-human-settlements-2011-cities-and-climate-change

43.Houghton A & Carlos C-S. Health co-benefits of green building design strategies and community resilience to urban flooding: A systematic review of the evidence. MDPI. 2017 Dec 06. https://www.mdpi.com/1660-4601/14/12/1519

44.He BJ. Green building: A comprehensive solution to urban heat. Energy and Buildings. 2022 Sep 15; 271:112306. https://www.sciencedirect.com/science/article/pii/S0378778822004777

45.Sharifi A, Pathak M, Jishi C & He BJ. A systematic review of the health co-benefits of urban climate change adaptation. Sustainable Cities and Society. 2021 Nov 01; 74:103190. https://www.sciencedirect.com/science/article/abs/pii/S2210670721004686

46.Abubakar IR & Dano UL. Sustainable urban planning strategies for mitigating climate change in Saudi Arabia. Environment, Development and Sustainability. 2019 Jul 16. https://link.springer.com/article/10.1007/s10668-019-00417-1

47.Chen WY. The role of urban green infrastructure in offsetting carbon emissions in 35 major Chinese cities: A nationwide estimate. Cities. 2015 Apr 01; 44:112–20. https://www.sciencedirect.com/science/article/pii/S0264275115000153

48.Younger M, Morrow-Almeida HR, Vindigni SM & Dannenberg AL. The built environment, climate change, and health: opportunities for co-benefits. American Journal of Preventive Medicine. 2008 Nov; 35(5):517–526. https://pubmed.ncbi.nlm.nih.gov/18929978/

49.Abuwaer N, Ullah S & Sai G A-G. Building climate resilience through urban planning: Strategies, challenges, and opportunities. Research Gate. 2024 Jan. https://www.researchgate.net/publication/376592150_Building_Climate_Resilience_Through_Urban_Planning_Strategies_Challenges_and_Opportunities

50.Adesina OS, Ayetan O, Otokiti KV & Ojotu O. Harnessing nature to address climate change: Agri-environmental approaches for adaptation and mitigation. In: Leal Filho W, Nagy GJ & Ayal DY, editors. Handbook of nature-based solutions to mitigation and adaptation to climate change. Springer International Publishing; 2024 Mar 31; 1–15. https://doi.org/10.1007/978-3-030-98067-2_95-1

51.Meerow S. Double exposure, infrastructure planning, and urban climate resilience in coastal megacities: A case study of Manila. Arizona State University. 2017 Nov 01. https://asu.elsevierpure.com/en/publications/double-exposure-infrastructure-planning-and-urban-climate-resilie

52.Bhargava A, Lakmini S & Bhargava S. Urban heat island effect: It's relevance in urban planning. Journal of Biodiversity and Endangered Species. 2017 Jan. https://www.researchgate.net/publication/318482561_Urban_Heat_Island_Effect_It's_Relevance_in_Urban_Planning

4

A Resilient Earth for Human Health

Abstract

The way we treat ourselves is a result of the way we have been treating everything that surrounds us. This reality has impacted non-human and human species alike. If we want to change that, we need to understand our role and how important biodiversity is to both human and planetary health. Our Earth's health is a reflection of our human health and mental wellbeing. Because of lack of movement and a sedentary lifestyle, we have lost our connection to forests, mountains, rivers, and oceans. If we go back to nature and understand our role as stewards, maybe, and just maybe, we will be able to take care of our health.

Keywords: Biophilia, planetary health, human health, mental wellbeing, environmental stewardship, biodiversity, ecosystem services, spiritual ecology, ecological resilience, and food security

Authors

Vera Urtaza Reyes, Co-Founder and Executive Director, Keystone Species Alliance [vera@keystonespeciesalliance.org]

Pooja Sharma, Director of Legal Advocacy and Policy, Keystone Species Alliance [pooja@keystonespeciesalliance.org]

Introduction

For decades, experts and scientists have warned that we cannot continue developing using the same extractive and unsustainable models. These models have resulted in an incremental impact on the natural cycles that sustain life on Earth and on the processes that will allow our societies a chance to adapt to the daunting circumstances that will come in the near future due to increasing global temperature.

This model has led to unsustainable practices such as overconsumption, land-use change, industrial agriculture, expanding urbanization, poor waste management, and pollution, to mention a few, which are now affecting the integrity and function of ecosystems, a critical foundation of the wellbeing of all life on earth.

Many scientists assert that we are now facing the Sixth Mass Extinction (and the only anthropogenic mass extinction). This is a profound but alarming statement that demands our serious attention. This potential extinction event results from extensive and intensive human activities that have dramatically altered our planet. As a part of biodiversity, humans have lost their way in stewarding natural cycles and protecting life on Earth.

What does this mean to us and to the world that we have created? If the Sixth Mass Extinction occurs, life as we know it, will undergo dramatic change. Our health and survival are intrinsically linked to the health of biodiversity which includes a rich variety of flora and fauna. Humans depend on biodiversity for food security, water, resilience, climate change mitigation and adaptation, regenerative economies, and peace.

The reality that one million species are teetering on the brink of extinction can only mean that humanity is also at risk. From pollinators to sharks, every species plays a crucial role in sustaining life on Earth. Through active engagement with our planet, we can redirect the disruptive processes that we have created to be more aligned with the ecological processes through which the web of life is created, sustained, and re-newed.

These risks are exacerbated by expanding urbanization, unsus-tainable food production and consumption patterns including increasingly complex food chains, poor waste management and disposal, increased trade and travel, as well as pollution, re-sulting in biodiversity loss, ocean acidification, desertification, and the climate crisis.

Our current way of life has severed our coexistence with nature. We no longer grow our food and often remain unaware of its origins and the processes it undergoes from the farm to our plates. This phenomenon has affected not only our relationship with nature but also the way we consume. This disconnect has not only damaged our relationship with nature, but has also led to excessive food waste, estimated to account for up to 10 percent of global greenhouse gas emissions (1). Food waste has a direct impact on our health as it exacerbates the direct impacts of climate change such as extreme heat waves and threatens food security. We, as individuals, have the power to change this. We go to grocery stores as an errand to run, taking in food that seems to "appear" without any respect for it. Overconsumption, along with the excessive food waste, is a recipe for cyclical harm.

This separation from nature also affects how we relate to one another and to ourselves, leading to a loss of intimacy and connection. Many people feel lost in a performative society, and our mental health suffers as a result. The pervasive sense of emptiness, rather than joy and self-realization, characterizes the modern era of depression.

The current degradation of biodiversity and ecosystems mirrors the state of human physical and mental health. Neglecting our wellbeing leads us to disregard our role as stewards of nature, focusing instead on productivity and self-worth measured by material success. For millennia, human development has prioritized civilization's needs over those of the planet, resulting in an extractive model that first harmed Mother Earth and now harms us This materialistic lifestyle has diverted us from our spiritual paths. We are spiritual beings as life itself, the miracle of experiencing our mere existence, is spiritual.

This mindset views the world as a machine and nature as dead matter. The polarity is distinct as we view nature as separate from us and, thus, offers an opportunity to conquer. Colonial narratives have historically promoted conquering and extraction when differences are present. The differences are seen as a conquering opportunity to take, extract, and exploit. These narratives have underpinned the business of capitalism as it exists today. As we see it, capitalism and colonialism are concerningly growing more similar in their ethos and general conduct.

This mentality that separates humans from the natural world has led to the modification of 14.6 percent or 18.5 million km^2 of the Earth (2). The direct consequence of these impacted ecosystems is that they have been degraded to a point where their ecosystem services are at risk. Just like the state of ecosystems, human civilization is now facing hunger and a variety of diseases related to how we produce, what we eat, and our sedentary lifestyle.

Despite this disconnect, we still have the potential to reconnect with nature if we look within ourselves. Re-establishing these connections requires time and effort for feeling the sun, walking barefoot, meditating, and appreciating the natural world around us. Recognizing the natural cycles that sustain life can foster this reconnection.

So, how do we reconnect to reshape and rewind our world? How do we slow down when the world demands that we identify with our productivity and wealth? When we update how we speak to ourselves and to other human beings, we reprogram our connection to ourselves, to one another, and to the land. In that active engagement of storytelling, we are reminded of our resilient qualities as humans. In this, we are a part of biodiversity. When we make active efforts to preserve biodiversity and mitigate biodiversity loss, we promote regenerative standards of living because we appreciate that biodiversity is an essential component of nature, critical to maintaining a functioning and resilient natural system that can continue to provide services for society and that we can do the same for Her, to live in a symbiotic way with nature (3).

One Health

The economic model in place has brought many benefits and substantial improvements to many. Unfortunately, it has come at the expense of nature. It has resulted in unsustainable consumption and production patterns that have taken a toll on nature's ability to replenish its resources. As mentioned before, the state of nature globally is not able to stand against these unsustainable and destructive ways of living. With human populations expected to continue growing, measures that consider the system on which life depends are much needed.

The Convention on Biological Diversity (CBD), World Health Organization (WHO), the Intergovernmental Science-Policy Platform on Biodiversity and Ecosystem Services (IPBES), and the UN Human Rights Council have collectively recognized biodiversity–health interlinkages (4). Indigenous peoples tend to approach health as an equilibrium of spirituality, traditional medicine, biodiversity, and the interconnectedness of all that exists. This leads to an understanding of humanity in a significantly different manner than for non-indigenous peoples (5). The concept of One Health recognizes these interdependencies and promoted an integrated, unifying approach that sustainably balances and optimizes the health of humans, animals, plants, and ecosystems (6). The OECD finds that global biodiversity finance is estimated at USD 78–91 billion per year (data from 2015–2017) (7). In contrast, health spending is significantly higher. In 2020, global spending on health reached US $ 9 trillion or 10.8 percent of global gross domestic product (GDP).

Marginalized populations are more likely to face elevated health risks from environmental change. These include lower-income communities and indigenous communities that are coping with environmental changes driven largely by economic processes in other parts of the world. They are especially vulnerable to disease risk as a result of multiple stresses. They have few resources for combating global environmental change and have little voice in the decision-making processes of local, regional, national or global policy institutions. Because health is a central element in sustainable development, poor communities face a double challenge. They are at greater risk of environmental health impacts which worsen the development challenges they face, and which in turn further weaken their ability to respond to health risks.

The Earth needs urgent action. With vulnerable communities at the front line of biodiversity loss and climate change, the threats caused by land-use change, unsustainable agricultural production and intensification, large-scale deforestation, land degradation, and biodiversity loss posed to the web of life are unsustainable and can create irreparable damage to this planet.

The health of ecosystems is the bedrock of human health. The intricate web of biodiversity that sustains these ecosystems is not just a part of our world. It is our world. Recognizing this interconnectedness is crucial for securing a just and equitable present and future for all life on Earth. A change in paradigm is not needed. It is our collective responsibility. Our failure to recognize what has been readily apparent to the Indigenous people is doubly shocking as we are ignoring wisdom for temporary solutions.

A global effort holds at its core acknowledgement of the large benefits that well-functioning systems provide. There is an urgency to understand the vital role played by biodiversity to prevent the spread of zoonotic and vector-borne diseases that could become epidemic or pandemic. As the International Union for Conservation of Nature (IUCN) noted in its November 2018 policy scoping brief at the 14th Conference of the Parties, "the use of systematic risk analyzed vulnerability assessments and integrated impact and strategic assessments to proactively manage non-communicable and infectious disease risks associated with biodiversity change, wildlife trade, and other emergence of disease."

At the same time, there is also an emphasis on "investment in nature-based solutions, such as the integration of bio-diverse green and blue spaces in urban development, improving availability of and accessibility to diverse diets, tightening control and rationalizing use of antimicrobial agents, pesticides and other biocides, and maximizing the health benefits of exposure to bio-diverse environments (8)." In 2018, the IUCN recommended that enhancing the environment to address health matters and identifying health stakeholders should be a priority in the design of conservation and restoration policies (8). Organizing and governmental bodies need to create working groups in which representatives from health and biodiversity are represented.

The loss of healthy ecosystems are risk factors that need to be addressed in a timely manner, as the impacts to human health due the increment of diseases that can be transmitted across the animal–human–plant, hold far reaching consequences to society, economic trade, food security, ecosystem function, and the health and wellbeing of humans (6). These events are expected to increase in time and will cause more damage due to climate change and biodiversity loss.

Intensive human activities have impacted nature in many ways. Urgent measures are needed to restore and conserve it. As a joint effort to present a broad and comprehensive framework to face these challenges, the United Nations through the Food and Agriculture Organization (FAO), the United Nations Environment Programme (UNEP), the World Organization for Animal Health (WOAH), and the World Health Organization (WHO), have developed a concept that englobes a unifying approach that aims to sustainably balance and optimize the health of people, animals, and ecosystems: One health.

This framework recognizes that the health of humans, domestic and wild animals, plants, and the wider environment (including ecosystems) is closely linked and interdependent (9). The goal is to address common threats to the health of humans, animals, and plants by promoting environmental and biodiversity protection through collaboration of all sectors. This interdisciplinary approach allows for clear tracking of cause and effect for zootrophic diseases. For example, in-depth understanding of ecological processes showed that mercury poisoning of fish and impending health risks for humans in the Amazon basin were not due to upstream gold mining but due to soil erosion following deforestation (10).

The loss of traditional knowledge through the displacement of indigenous cultures and the loss of species through land use change and overharvesting continues to pose a significant threat to people's health and wellbeing. The loss of intellectual property rights remains problematic for many indigenous cultures and arises not only through the transfer of traditional knowledge, innovation, and practices in the public domain but also through unauthorized access to and appropriation of such knowledge.

Holistically, it bridges humans, animals, and the environment to address the full spectrum of disease control from prevention to detection, preparedness, response, and management, and contribute to global health security (9). The One Health approach reminds us of our interconnectivity to science, to ecology, and to one other.

We shaped the world

In nature, certain non-human species play crucial roles in shaping, sustaining, and balancing ecosystems. These are known as keystone species. They are divided into three categories: apex predators, mutualists, and ecosystem engineers. Without keystone species, ecosystems would either collapse or change drastically. The role of keystone species is integral in that they balance the habitat's trophic cycle simply by existing and functioning, thus playing a fundamental role in many ecosystem services such as sequestering carbon and bringing health to our food systems, among others. With our planet containing ecosystems that function as carbon sinks, keystone species are critical to the functioning of carbon sinks in both land and water. Have we wondered, as humans, if we can be empowered to balance our existing ecosystem through our existence and functionality? (11)

Humans are keystone species. Yet, we have lost the ability to function harmoniously. Despite possessing all the characteristics of a keystone species, our actions have reshaped the world in a way that has impacted every corner of the Earth, leading to the extinction of numerous species. As our population grew, human-made mass, referred to as the "anthropogenic mass," surpassed global living biomass, infiltrating even the deepest parts of the ocean with plastic waste (12). The health of life below water is crucial for a healthy Earth.

Although we often think of ourselves as separate from the natural world, we are, in fact, part of it. From the moment we began growing food and forming settlements, we started shaping our environment. Indigenous knowledge systems have long recognized the necessity of protecting nature through a symbiotic, respectful, and reciprocal relationships.

As settlements grew into cities, this transformation catalyzed exploitative changes over the natural world. Cities which now house 4.4 billion people, a number projected to surpass six billion by 2050, consume vast natural resources and produce significant waste. Covering only 2-3 percent of land, cities account for 75 percent of natural resource consumption, up to 80 percent of energy consumption, 70 percent of greenhouse gas emissions, and 50 percent of waste production (13).

The growth of cities puts ecosystems at risk. To break away from the narrative of separation from these ecosystems, we must recognize dependence on them for air and water purification, pollination, food security, and climate change mitigation. Climate change impacts, such as extreme weather, directly affect our health through respiratory diseases, heat-related illnesses, and food scarcity. Further, the lack of connection to nature, the absence of green and blue spaces, and the monotony of urban environments adversely impacts our mental health. Increasing mental health problems including depression are prevalent among city dwellers (14).

There is evidence to show that nature, biodiversity, and green and blue spaces improve public health through nature-based solutions. Exposure to nature has enormous benefits for mental and physical health. Both green and blue spaces have been linked to reductions in cardiovascular diseases and symptoms of anxiety and depression.

The concept of "biophilia" suggests that humans have an intimate emotional attachment to other humans and to nature, especially living biota, and that because humans evolved within nature, we still display inherited adaptations, making us likely to function well when exposed to natural environments (14). Biophilia connect us to ourselves, our surroundings, and nature. Cities around the world are implementing nature-based solutions to mitigate and adapt to climate change and to reduce health-related issues.

Nature in cities can improve air quality, strengthen immune systems, and enhance overall wellbeing. Effective actions should focus on creating healthy cities and environments. It really is that simple. Biodiversity also benefits from these initiatives as many species can adapt to urban environments making well-managed urban areas into healthy ecosystems. Integrating nature education can improve how future generations relate to and protect their environments. George Monbiot´s essay "Rewild the Child '' highlights that children who spend more time in natural environments perform better academically and develop stronger leadership qualities. His findings underscore the long-term benefits of nature exposure for children, including better memory, attention, and social skills (15).

Reconnecting with nature provides peace and grounding, countering the anxiety induced by modern economic models. Preventive measures to secure access to food, clean water, medicines, and air are urgent and protecting and restoring biodiversity is our only chance. Biodiversity sustains ecosystems, which, in turn, support all life.

Cities, despite covering a small land area, have massive environmental impacts due to their high consumption and waste production. They are built with boundaries that often ignore natural geographic markers, reinforcing a disconnect from nature. The concept of bioregionalism, which integrates natural boundaries into territorial markers, is gaining traction as a way to restore urban biodiversity and create resilient cities that act as carbon sinks.

Healthy ecosystems are crucial for combating biodiversity loss and climate change. Natural carbon sinks, found in both land and oceans, absorb more than half of all greenhouse gas emissions, preventing more rapid warming. Conserving and restoring natural spaces are essential for limiting carbon emissions and achieving necessary mitigation in the coming decade.

Taking better care of our surroundings can lead to better self-care. Restoring cities to be more nature-based can inspire people to adopt more active and connected lifestyles, becoming better stewards of their environment. Many feel powerless in the face of environmental challenges, but bringing nature back to cities can help people become more aware of their impact on ecosystems. This awareness, in turn, can improve human health globally.

Cities built with fences and walls act as gridlocked borders where human citizens unquestioningly abide by designated boundaries. Our perception of boundaries is beholden to concrete artificial mechanisms instead of the Earth's geographical markers such as rivers, mountains, and oceans. Our systems are dated and represent a time when borders were created to reign sovereignty within political lines.

Bioregionalism is not new but has recently gained more traction as we understand the importance of integrating nature's natural lines as territory markers. By restoring urban biodiversity and creating resilient cities that act as carbon sinks and sanctuaries for rich flora and fauna, we can positively impact human health, both physically and mentally.

We ate the world

The Holocene epoch brought stable weather conditions that enabled the emergence of agriculture and human settlements. For approximately ten thousand years humans farmed ecologically considering nature's systems and cycles. This period allowed for a diverse diet, thriving biodiversity, and a balanced climate. Early agricultural systems had a lower environmental impact, maintaining harmony with nature. However, as human populations and cities grew, so did the demand for food, leading to the birth of industrial agriculture. This system views nature as a commodity requiring human intervention rather than a network of interconnected processes.

Industrial agriculture's reliance on chemical fertilizers has degraded soil health, leading to the loss of 24 billion tonnes of fertile soil annually (16). Healthy soil rich in microorganisms retains water, resists erosion, and stores more than 4,000 billion tonnes of carbon (17). The loss of healthy soils threatens food security for 1.5 billion people and disrupts ecosystems, contributing to biodiversity loss and climate instability. Soil, once seen as a living entity teaming with microorganisms, is now treated as inert dirt needing chemical fertilizers. However, soil health is crucial for sustaining life as it houses billions of organisms that support the food web and ecological processes.

Healthy soil microbiomes are essential for nutrient cycling and plant health. Pesticides, herbicides, and fungicides disrupt these microbiomes, reducing the nutritional quality of food and harming human health. There are over 1,000 pesticides in use today, many of which pose significant risks to humans and the environment. Pesticides have been linked to numerous health issues including cancer, obesity, neurodegenerative diseases, and developmental disorders. They also harm non-target species including pollinators like bees which are crucial for food production. A 2017 UN report highlighted the dangers of excessive pesticide use and called for safer, more sustainable agricultural practices (18).

Industrial agriculture has not fulfilled its promise of feeding the world. The latest State of Food Security and Nutrition in the World (SOFI) report reveals that 2.4 billion people faced food insecurity in 2021 with 900 million experiencing severe food insecurity (19). Additionally, 3.1 billion people could not afford a healthy diet (19). Industrial agriculture's focus on monocultures and chemical inputs has diminished the nutritional value of food, contributing to malnutrition and health problems. Historically, human diets were diverse and nutrient-rich with over seven thousand species used for food. Monocultures and industrial agriculture have reduced this diversity impacting health and cultural heritage. Diverse, locally produced diets are essential for nutrition and cultural identity.

The benefits of having a diverse local diet are innumerable. Diversity means nourishment. Our diets used to be diverse full of nutrients, colors, and shapes, and were produced locally. More than seven thousand species have fed humanity throughout

87

history (16). The more diverse our plates, the more-dense in nutrients they will be. Food is medicine.

Monocultures have impacted ecosystems and our culture. For many people, food is the glue that holds families together, and brings back memories from childhood, and comforts them. Monocultures have reduced access to these nourishing dishes that have formed part of our heritage and identity. Mexico and many other countries are clear examples of how food diversity and culture are bonded together. Mexican cuisine is categorized as Intangible World Heritage by United Nations Educational, Scientific and Cultural Organization (UNESCO) and each region can find its history in every dish.

Due to the intensification of land use that industrial agriculture has had, we are now facing food insecurity, but we are also facing the loss of a part of us. With the increase in global temperatures, we will need to adapt what we plant so that the seed can survive environments and inhospitable conditions. Diets will change, and so will culture. Our food system is the third cause of all greenhouse gasses. Livestock is responsible for at least two-thirds of the damage done and for contributing to deforestation, biodiversity loss, competition for edible grains, and poor animal welfare conditions (20).

Livestock production, particularly beef, is a major driver of deforestation, especially in tropical regions. This deforestation contributes to biodiversity loss, carbon emissions, and the degradation of ecosystems vital for climate regulation. Sustainable livestock practices, such as silvopastoral systems, can help mitigate these impacts. Nitrogen is vital for plant growth and living soils contain nitrogen-fixing bacteria that support this process. Industrial fertilizers disrupt these natural processes leading to soil degradation and dependence on chemical inputs. Synthetic fertilizers, derived from nonrenewable sources, contribute significantly to greenhouse gas emissions.

Furthermore, the current rates at which our world is changing from land-use change and climate change, our current economic development practices are transforming the interfaces between wildlife, livestock, and humans, increasing the incidence of disease emergence and spread (21)(22). The overuse of antibiotics in livestock farming poses a serious threat to human health by promoting antibiotic resistance. The WHO recommends limiting antibiotic use in animals to preserve their effectiveness for human medicine. Sustainable livestock practices can reduce the need for antibiotics and improve animal welfare. To address these risks, land managers and policy-makers at various levels are incorporating the One Health approach into their environmental and agricultural policies, emphasizing the need to address the interconnections between human health, animal health, and ecosystem health (23). The IUCN World Commission on Protected Areas recommends that this approach should include not only agricultural and livestock areas but also protected and conserved areas, as they play a crucial role in the wildlife-livestock-human interface. This integration is essential for minimizing, mediating, and monitoring infectious disease risks (24).

Organic farming, rooted in ancestral knowledge, promotes nutrient-rich food production in harmony with natural cycles. It supports both human and environmental health by maintaining soil fertility and ecological balance. Organic farming methods enhance biodiversity, sequester carbon, and reduce the need for harmful pesticides and fertilizers. Agroecology promotes farming practices that work with natural processes enhancing biodiversity, soil health, and ecosystem services. This approach reduces the need for chemical inputs, supports pollinators, and improves food security. Agroecology recognizes the interconnectedness of human and environmental health, offering a sustainable path forward. Pollinators, such as bees, are essential for the reproduction of 80 percent of flowering plants and contribute significantly to global food production. Protecting pollinators from pesticides and habitat loss is crucial for food security and ecosystem health. Biodiversity supports the discovery of natural medicines and helps regulate diseases. Healthy ecosystems with diverse species can reduce the spread of zoonotic diseases, highlighting the importance of conservation for public health.

By promoting sustainable agricultural practices such as organic farming and agroecology, and enhancing urban ecosystems, we can mitigate the impacts of climate change and biodiversity loss. This holistic approach not only aims to reduce the emergence and spread of diseases but also to foster a healthier planet where all forms of life can thrive. Restoring ecosystems enhances their ability to sequester carbon and buffer against extreme weather events. Forests, wetlands, and other ecosystems play a vital role in climate adaptation, protecting infrastructure and livelihoods from climate impacts. The current industrial agricultural system is unsustainable. It is harming both the environment and human health. A shift towards organic farming and agroecology can restore soil health, enhance biodiversity, and improve food security. By recognizing the intrinsic value of nature and working in harmony with natural processes, we can create a more sustainable and resilient food system that supports human and planetary health.

Rewind to rebuild

The planet's wellbeing, people's health, and society's stability are interconnected. In the way we have developed, we have not taken care of the web of life that provides the means for us to live in a healthy, balanced, and harmonious way with all life on Earth. This web of life is vital for us to secure a livable planet in the future. By caring for what surrounds us and acquiring consciousness of our impact on the ecosystems that provide oxygen, water, and food, we also care for ourselves as a global community.

Despite the narrative behind "developed societies," we are now witnessing one billion people suffering from hunger and malnutrition, two billion suffering from diseases like obesity and diabetes, and countless others who have cancer and pollution-related diseases. If we aim to change this, we need to change the paradigm of exploitation for both humans and the natural world to improve mental and physical health. We are nature and need to start seeing ourselves as part of the planet Earth's delicate system.

Moreover, even if we do not find ourselves aligned with this mindset, rough numbers show that consumerism and industrial agriculture are depleting life on Earth and not satisfying people's basic needs. Industrial agriculture accounts for 75 percent of the ecological damage to the planet, but it does not feed everyone. It has impacted every corner of the planet and only benefited the ones behind it. Our economies depend on a stable climate and thriving biodiversity; species make the fabric of healthy ecosystems. Healthy ecosystems provide 17 ecosystem services for the entire biosphere; the value (most of which is outside the market) is estimated to be US $16-54 trillion annually, with an average of US $33 trillion annually (25).

There is an urgency to recognize that we need to replace this system with one that can nourish all species while protecting water, biodiversity, and soil fertility. There is a global call to recognize how dependent we are on nature and that by protecting nature, we are also protecting livelihoods, the economy, human rights, health, democracy, and peace. We can choose to step away from an industry that is poisoning the planet, diminishing planetary health, and is profit-driven. In that case, we will achieve substantial changes that will help us all to have a better quality of life.

In our nature, we can find the means to start appreciating that the resources vital to the maintenance of life like biodiversity, soils, and water, are sacred and not "commons" meant to be exploited. Through healing and compassion, we can now identify ourselves as co-creators of Mother Earth. If we see the state of nature as a reflection of human health, we can see a clear correlation between how we neglected our place as stewards by neglecting our health. The more we grew, the more we became disconnected from our surroundings, the more we consumed, the more we lost ecosystems. The facts can be physical, but we must also consider what will become of us if we continue on this path.

Many naturalists have argued that the extermination of species represents a spiritual and intellectual impoverishment for humanity. A world without other earthly companions would not merely be a more dangerous place, it would also be much lonelier and more desolate. What will become of the human spirit when the inspirited creatures we have invoked over millennia in our most enlightened cultural traditions are gone?

While there is dark matter permeating our world through their extractive methods, the light counteracts the taking with resilience and effective impact-based action. Small to large sized organizations are playing an unignorable role in pushing strong conservation measures forward. For example, Keystone Species Alliance, Earth Law Center, and the Center for Whale Research partnered to end extractive logging measures in the Elwha Forest located in the state of Washington, United States. The movement lawyering behind the advocacy efforts consisted of:

1. Rights of nature litigation, which included litigating on the behalf of Chinook Salmon and Southern Resident Orcas – two non-human species that are keystone species and integral to the health of the Elwha River.
2. Garnering support to buy the impacted land so it remains protected at all costs.
3. Twenty-minute impact documentary featuring key stake-holders and drivers behind the litigation and Elwha campaign.

4. Ongoing education series of protecting the western red cedars as cultural keystone species.

The above highlighted examples are a snapshot of what can happen when organizations come together to advocate for identifying, protecting, and restoring keystone species as recognized markers of a healthy habitat, and thus a healthy food system (26).

Similarly, the power of data gathering from field work pushes legislation forward in an impactful, tangible way. The Keystone Species Alliance's fellowship program, which invites on the ground researchers to study and monitor keystone species in their respective habitat, is linked to building resilience in the local community through education of the respective keystone species' significance and the importance to protect and restore their populations. Why is this important, and how is it linked to better population health? When we track the health of keystone species in their natural habitats to assess their behavior, population size, migratory patterns, and survival rate, we can make conclusive findings about which habitats need deeper protection. Unsurprisingly so, the habitats that need deeper protection are rich in biodiversity and contemplative of securing a deeper relationship with oneself and nature.

The Keystone Species Fellowship Program's first fellow, Callie Veelenturf of The Leatherback Project, studies leatherback sea turtles in their critical habitat of threatened populations to understand local perspectives and relationship between the Indigenous and black communities with sea turtles in the

Darien gap. Our reliance on a healthy ocean is an incontestable fact. She found that examining migratory patterns and fishery interactions of this powerful marine aquatic keystone species helps us determine the health of our ocean wherein we, as humans, need to take action (26). In humans taking action for their non-human counterparts, they are taking action for themselves. For example, as identified by the World Health Organization, investing in nature-based solutions such as the integration of biodiverse green spaces in urban development, improving availability of and accessibility to diverse diets, tightening control and rationalizing use of antimicrobial agents, pesticides and other biocides, maximizing the health benefits of exposure to biodiverse environments, and better monitoring of environmental change in line with the "One Health" approach (27). Without understanding the true connectivity between ourselves and nature, we suffer in our inaction. In 2015, WHO and the Secretariat of the Convention on Biological Diversity signed a Memorandum of Understanding to strengthen collaboration and, with relevant partners, capitalize on their respective scientific and technical expertise on the links between health and biodiversity (27). Almost a decade has passed and we are still learning from our own mistakes. A global pandemic occurred where flora and fauna flourished in bounty, creating natural green spaces we all innately gravitated towards. In the collective suffering of the unknown, there was a childlike awe and elatedness with wildlife populations regrowing, air clearing, and lushness thriving.

Merriam-Webster defines rewinding as "an effort to increase biodiversity and restore the natural processes of an ecosystem that typically involves reducing or ceasing human activity and often the planned reintroduction of a plant or animal species and especially a *keystone species*" (28). In order for us to rebuild the world we have harmed, we must look at what will create a lasting impact with measurable results. On-the-ground efforts to build stronger, resilient communities where the environment and its non-human species play a leading role is necessary for rebuilding our world. We can go from sustaining our current modus operandi and shift towards a regenerative model where our habitats are prioritized not for what they provide to us but for what we can offer to the habitat.

Concluding comments

The wellbeing of our planet, the health of people, and the stability of society are profoundly interconnected. Our current development trajectory has often disregarded the intricate web of life that sustains us, which is crucial for securing a livable planet for future generations. By becoming more conscious of our impact on the ecosystems that provide oxygen, water, and food, we can better care for ourselves as a global community. Despite the progress associated with "developed societies," we now face severe consequences: one billion people suffer from hunger and malnutrition, two billion grapple with diseases like obesity and diabetes, and countless others are afflicted by cancer and pollution-related ailments. To reverse this trend, we must shift from a paradigm of exploitation to one that fosters mental and physical health for both humans and the natural world. Recognizing our integral role within Earth's delicate system is essential.

Understanding the sacredness of biodiversity, soils, and water is vital for sustaining life. Through healing and compassion, we can identify ourselves as co-creators of Mother Earth. Even if not everyone aligns with this mindset, the evidence shows that consumerism and industrial agriculture are depleting Earth's resources without meeting basic human needs. Industrial agriculture is responsible for 75 percent of ecological damage yet fails to feed everyone, benefiting only a few. Organic farming and agroecology offer a way to create a food system that respects the interconnectedness of life. These approaches promote regenerative agriculture and enhance ecological services, acknowledging Earth as a living system. Moving away from fossil fuel dependency and minimizing the use of "energy slaves" is crucial in combating climate change and biodiversity loss.

Our economies rely on a stable climate and thriving biodiversity. Healthy ecosystems provide 17 essential services valued at US $16-54 trillion annually, with an average of US $33 trillion (25). The state of nature reflects human health. Neglecting our role as stewards has led to both ecological and health crises. As we grew and consumed more, we disconnected from our surroundings, resulting in significant ecosystem loss. Developing and implementing strategies focused on restoring urban ecosystems can lead to significant improvements in human health, climate change mitigation, and biodiversity preservation. Initiatives such as creating greenways, urban wetlands, and green infrastructure, alongside protecting native species and remaining ecosystems, can have a profound spillover effect.

We must replace the current profit-driven, environmentally

damaging system with one that nourishes all species while protecting water, biodiversity, and soil fertility. This global call recognizes our dependence on nature and emphasizes that protecting it safeguards livelihoods, the economy, human rights, health, democracy, and peace. By stepping away from industries that poison the planet, we can achieve substantial changes and enhance quality of life for all. If we see the state of nature as a reflection of human health, we can see a clear correlation between how we neglected our place as stewards by neglecting our health. The more we grew, the more we became disconnected from our surroundings, the more we consumed, the more we lost ecosystems. The facts can be physical, but we must also consider what will become of us if we continue on this path. Many naturalists have argued that the extermination of species represents a spiritual and intellectual impoverishment for humanity. A world without other earthly companions would not merely be a more dangerous place, it would also be much lonelier and more desolate (29). What will become of the human spirit when the inspirited creatures we have invoked over millennia in our most enlightened cultural traditions are gone? (29)

In recognizing the interconnectedness of all life and the critical role we play within Earth's ecosystems; we can shift our development trajectory towards sustainability and harmony with nature. By embracing practices that restore and protect our environment, we can ensure a healthier, more stable, and more fulfilling existence for ourselves and future generations. The time to act is now; the future of our planet and our own wellbeing depends on it.

References

1.Environment UN. UNEP food waste index report 2021. UNEP - UN Environment Programme. 2021. https://www.unep.org/resources/report/unep-food-waste-index-report-2021

2.Routley N. Mapped: Human impact on the Earth's surface. Visual Capitalist. 2021 Oct 15. https://www.visualcapitalist.com/mapped-human-impact-on-the-earths-surface/

3.PRI. Nature in responsible investments: Introductory guide. Principles for Responsible Investment. 2024 Mar 25. https://www.unpri.org/sustainability-issues/nature-in-responsible-investments/12149.article

4.Willets L, Siege C, Anna M, Horn O, et. al. Advancing integrated governance for health through national biodiversity strategies and action plans. Lancet. 2023 Sep 02. https://pubmed.ncbi.nlm.nih.gov/37499672/

5.World Health Organization. Global plan of action for health of indigenous peoples. World Health Organization. https://www.who.int/initiatives/global-plan-of-action-for-health-of-indigenous-peoples

6.FAO. One Health Joint Plan of Action, 2022–2026. Working together for the health of humans, animals, plants and the environment. FAO; UNEP; WHO; World Organisation for Animal Health (WOAH) (founded as OIE). 2022. http://www.fao.org/documents/card/en/c/cc2289en

7.OECD. A comprehensive overview of global biodiversity finance. Organisation for Economic Cooperation and Development. 2020 Apr. https://www.oecd.org/content/dam/oecd/en/publications/reports/2020/04/a-comprehensive-overview-of-global-biodiversity-finance_ad660ace/25f9919e-en.pdf

8.14th Conference of the Parties & Sheik S. IUCN's policy briefing paper on biodiversity and human health and well-being. IUCN. 2018 Nov 17-29. https://view.officeapps.live.com/op/view.aspx?src=https%3A%2F%2Fportals.iucn.org%2Flibrary%2Fsites%2Flibrary%2Ffiles%2Fresrecrepattach%2FDRAFT%2520IUCN%2520Policy%2520Briefing%2520Paper%2520on%2520Biodiversity%2520and%2520Human%2520Health%2520CBD%2520COP14%2520051118.docx&wdOrigin=BROWSELINK

9.World Health Organization. One health. World Health Organization. https://www.who.int/health-topics/one-health

10.Forget G & Lebel J. An ecosystem approach to human health. International Journal of Occupational and Environmental Health. 2001 Apr/Jun. https://www.iai.int/admin/site/sites/default/files/uploads/International_Journal_An_Ecosystem_Approach_to_Human_Health.pdf

11.Aligned Magazine. Lyla June Johnston: Rewriting history & herstory. Aligned. https://www.alignedmag.com/people/lyla-june-johnston-rewriting-history-herstory/

12.Elhacham E, Liad B-U, Jonathan G, Yinon M & Milo R. Global human-made mass exceeds all living biomass. Nature. 2020 Dec 09. https://www.nature.com/articles/s41586-020-3010-5

13.IUCN. Cities and nature. IUCN. https://iucn.org/resources/issues-brief/cities-and-nature

14.Moran D & Turner J. Turning over a new leaf: The health-enabling capacities of nature contact in prison. Social Science and Medicine: Elsevier. 2019 Jun. https://www.sciencedirect.com/science/article/pii/S0277953618302752

15.George Monbiot. Rewild the child. Guardian. 2013 Oct 07. https://www.monbiot.com/2013/10/07/rewild-the-child/

16.Shiva V. Who really feeds the world? North Atlantic Books. https://www.northatlanticbooks.com/shop/who-really-feeds-the-world/

17.Global Agriculture. Soil fertility and erosion. Global Agriculture. https://www.globalagriculture.org/report-topics/soil-fertility-and-erosion.html

18. United Nations. Pesticides are "global human rights concern", say UN experts urging new treaty. OHCHR. 2017 Mar 07. https://www.ohchr.org/en/press-releases/2017/03/pesticides-are-global-human-rights-concern-say-un-experts-urging-new-treaty

19.World Health Organization. 122 million more people pushed into hunger since 2019 due to multiple crises, reveals UN report.

World Health Organization. 2023 Jul 12. https://www.who.int/ news/item/12-07-2023-122-million-more-people-pushed- into-hunger-since-2019-due-to-multiple-crises---reveals -un-report

20.Gonzalez VA. Can we raise livestock sustainably? A win-win solution for climate change, deforestation and biodiversity loss. The Conversation. 2022 Mar 14. http://theconversation.com/ can-we-raise-livestock-sustainably-a-win-win-solution-f or-climate-change-deforestation-and-biodiversity-loss-17 6416

21.Nova N, Athni TS, Childs ML, Mandle L & Mordecai EA. Global change and emerging infectious diseases. Annual Review of Resource Economics. 2022 Oct; 14:333–54. https://pmc.ncbi.n lm.nih.gov/articles/PMC10871673/

22.Jines KE, Patel NG, Storeygard A, Balk D, Gittleman JL & Daszak P. Global trends in emerging infectious diseases. Nature. 2008; 451:990-993. https://www.nature.com/articles/nature 06536

23.Plowright RK, Reaser JK, Locke H, Woodley SJ, Patz JA, Becker DJ, et al. Land use-induced spillover: a call to action to safeguard environmental, animal, and human health. Lancet Planet Health. 2021 Apr 06;5(4): e237–45. https://pmc.ncbi.n lm.nih.gov/articles/PMC7935684/

24.Hopkins SR, Olson SH, Fairbank HT, et. al. Protected areas and One health. PARKS. 2024 May. https://parksjournal.com/ wp-content/uploads/2024/05/PARKS-301-ALRE8783-Hopki ns-etal-1.pdf

25.Costanza R, Arge R, Groot R, Farber S, et. al. The value of the world's ecosystem services and natural capital. Nature. 1997 May 15. https://www.nature.com/articles/387253a0

26.Keystone Species Alliance. An alliance for all keystone species, the avenger heroes of our time, saving the Earth. They need human allies. Keystone Species Alliance. https://keyston especiesalliance.org/

27.World Health Organization. Health, environment and climate change. World Health Organization. https://www.who.i nt/news-room/fact-sheets/detail/climate-change-and-heal th

28. Merrian W. Rewilding definition & meaning. Merriam-Webster. https://www.merriam-webster.com/dictionary/rew ilding

29.Broswimmer FJ. Ecocide: A short history of the mass extinction of species. Pluto Books. 2015. http://www.jstor.org/ stable/10.2307/j.ctt18dzv1d

5

Water, Health, and Power: Challenges and Perspectives from Mexico City

Abstract

The author examines the international consensus on the Human Right to Water, Sustainable Development Goal 6 and its connection with human health using the standards set by the World Health Organization.

The situation of Mexico's water resources is addressed with an emphasis on the aquifer that supplies its capital city. A comparison is made with Mexican regulations for meeting the standards.

The prevailing situation in the aquifer where groundwater is being partially monopolized by concessions granted to local companies supplying the Transnational Water Bottling and Carbonated Beverage Companies is addressed.

The situation afflicting Mexico City is examined as water, which is perceived to be a scarce resource, is improperly managed. It is of questionable quality and is, therefore, impacting the health and economy of Mexicans.

Finally, the author reflects on perspectives regarding the commodification of drinking water and its impact on the full enjoyment of the human right to water.

Keywords: Human right to water, SDG 6, water quality, pollution, concessions, bottling companies, microbiological, chemical, radiological, acceptability, agroindustry, tap water, aquifer, and groundwater.

Author

Claudia Fernanda Padilla Rangel, Technical Secretary, Committees and Operations [cfpadill@gmail.com]

Human right to water, SDG 6, and WHO standards

Freshwater is a fundamental requirement of our daily lives. It stands as one of the nine planetary boundaries that must not be exceeded for humanity to operate safely.

The maintenance of the hydrological cycle is a crucial process that ensures the stability and functioning of the planet. Transgressing it leads to catastrophic consequences for humanity and for the maintenance of life (1). In this context, water and its management are alarming issues as water continues to be used at a rate that exceeds nature's capacity to regenerate it.

According to the latest World Water Development Report of 2023, water scarcity is present in large parts of the world along with accelerating freshwater pollution (2). Coupled with modifications to the hydrological cycle and climate change, this has adversely affected the availability of water for human consumption.

General Comment 15 of the International Covenant on Economic, Social and Cultural Rights (ICESCR) on the Human Right to Water, recognizes that water is a limited natural resource and a fundamental public good and that everyone is entitled to have sufficient, safe, acceptable, accessible, and affordable water for personal and domestic use (3). Additionally, the UN Resolution 64/292 states that "the right to safe and clean drinking water and sanitation, a human right is essential for the full enjoyment of life (4)".

In the same vein, Sustainable Development Goal 6 seeks to ensure the availability of water by promoting sustainable management and sanitation for all by monitoring 11 indicators. For this analysis, three indicators have been examined.

Monitoring conducted by the Government of Mexico indicates that:

1. 43 percent of Mexicans use a safely managed drinking water service.
2. There is 45 percent of water stress level (extraction of freshwater in proportion to available resources).
3. 55 percent of the monitored water bodies have good water quality (5).

The quality aspect of the human right to water entails the provision of safe water that does not contain microorganisms or chemical or radioactive substances. In other words, its physical, chemical, and biological properties are adequate with acceptable color, odor, and taste, making it wholesome (3).

By developing standards, the World Health Organization (WHO) ensures water quality for human consumption to protect public health. These standards are primarily intended for policy-makers to guide them in the development of national regulations (6).

The framework encompasses regulations and standards, safety plans, monitoring and management, and surveillance to ensure safe water supply. Table 1 shows the microbiological, chemical, and radiological pollutants, and their acceptable levels in water.

TABLE 1
MICROBIOLOGICAL, CHEMICAL, AND RADIOLOGICAL POLLUTANTS
AND THEIR ACCEPTABILITY BASED ON THE FOURTH WHO EDITION

MICROBIOLOGICAL	CHEMICAL	RADIOLOGICAL	ACCEPTABILITY
It includes bacteria, viruses, protozoa, and other organisms.	The criterion applied here is that the concentration of the chemical that does not pose any significant health risk when consumed over a lifetime. Only those with prolonged exposure represent a danger.	Presence of naturally occurring radionuclides.	This characteristic addresses the taste, odor, and appearance of water for human consumption.
This contamination is mainly from human and animal feces. This includes Guinea Worm (Dracunculus medinensis), and cyanobacteria, and Legionella.		Radon is also considered.	
		Levels in drinking water below which no action is required are 0.5 Bq/L for total alpha activity and 1 Bq/L for total beta activity.	Monitoring of these substances should be conducted when consumer complaints arise even though not all detected pose a health risk, as it may reveal some type or level of contamination.
Transmission routes include ingestion as well as inhalation, aspiration, and contact.	Those originating from industrial sources and inhabited areas, agricultural activities, and chemicals used for water treatment, and materials in contact with water and pesticides.		

Source: Compiled by the author based on the Guidelines for
Drinking-Water Quality:
Fourth Edition incorporating First Addendum by the WHO (6).

Water quality determined by these four parameters should be monitored by the State. This is crucial for ensuring compliance and implementing actions in order to improve water quality. Lack of documented and monitored information makes it difficult to take actions aimed at guaranteeing this right.

There is still a global challenge in reporting water quality data as there are different parameters and standards according to national realities. Nevertheless, the WHO guidelines and materials are considered elementary for ensuring a water secure future.

The achievement of SDG 6 and the human right to water are interconnected. They ensure health, human wellbeing, and environmental protection.

The importance of water quality for human health

The quality of both surface and groundwater depends on various factors including biogeochemical changes resulting from climate change, contamination, inefficient use, (industrial, agricultural, and domestic) where micropollutants, microplastics, and pharmaceutical substances, mostly from the chemical industry, play a role.

According to WHO guidelines regarding the microbiological aspect, it is important to prevent the entry of pathogens into water as they affect health. Examples include blood infections, lung disease, fever, joint pain, severe dehydration, anemia, vomiting, colic, diarrhea, muscle ailments, elevated heart rate, low blood pressure, gastroenteritis, conjunctivitis, and hepatitis A, among others (7).

This type of contamination can be avoided. Yet, unsafe drinking water is one of the factors that causes 1,500,000 deaths per year due to diarrhea which mostly affects infants and young children (8). Pregnant women, people living with HIV, and the elderly are also among these vulnerable groups.

Chemical contamination is caused by a wide variety of products of industrial origin. These include arsenic, copper, nitrate, and radon. These contaminants cause vomiting, diarrhea, damage to blood vessels, cancer, neurological disorders, and skin conditions, among others (9).

Arsenic is a naturally occurring chemical of the Earth's crust and is present in groundwater. Prolonged exposure to arsenic can cause conditions such as hyperkeratosis and pulmonary and cardiovascular diseases. The International Agency for Research on Cancer has classified arsenic and its compounds as carcinogenic agents (10).

It is estimated that between 94,000,000 and 220,000,000 people are potentially exposed to drinking water with elevated arsenic concentration. Prolonged exposure through drinking water and food can cause skin lesions. It is also associated with diabetes. Exposure to arsenic in utero and in early childhood can lead to poor cognitive development and increased mortality (10).

There has been a worrying spread of emerging chemical contaminants due, in part, to the limited understanding of their adverse health effects. These chemicals fall into three categories: pharmaceuticals chemicals, personal care products, and endocrine disruptors (11).

Radiological contaminants in water are found naturally. However, they have increased with economic, medical, nuclear and, above all, industrial activities such as mineral and hydrocarbon extraction. Radionuclides tend to disintegrate generating secondary products that are more radioactive such as uranium and radium which emit alpha and beta particles, and gamma rays (12).

Consumption of water with high levels of these radioactive particles can cause cancer, genetic defects, anemia, gastrointestinal disorders, osteoporosis, fetal malformation, cataract, bone growths, kidney disease, and liver disease. They can also compromise the immune system.

The spread of these health threatening chemicals is a manifestation of the impact of the prevailing development model wherein productive processes have been intensified at an accelerated and uncontrolled rate. In addition, exponential population growth has led to increased consumption. This can be quantified by observing the water footprint which measures the real demand for water. In other words, it quantifies virtual water.

In Mexico the last such study was conducted during 1996-2005. Mexico occupies the second place worldwide as a net importer of virtual water. It has an external water footprint of 42.5 percent and an internal (national) water footprint of 197,425 hm3/year distributed as follows:

- 92 percent agricultural sector (56% internal and 44% external)
- 3 percent industrial (33% internal and 67% external)
- 5 percent domestic (100% internal) (13)

Statistics reveal that there are several serious challenges in this area which force us to rethink public policies and management and also question and reformulate an export production model of virtual water where consumption increasingly needs additional territories since an exponential consumption model cannot continue with finite resources.

Water quality in Mexico City and its use by sector

Water quality is determined by optimal ecosystem functioning and integrated management. Monitoring of both these ensures a reliable supply of water, both groundwater and surface water. It also guarantees a supply. It is ideal to reduce the use of non-renewable resources such as the use of groundwater from aquifers whose recharge can take thousands of years to regenerate. We have to take care of the surface flow available in rivers and streams which, in turn, determines their self-regeneration.

In order to understand the quality of water in Mexico, we will detail its hydrological reality starting with its unfavorable natural distribution. In the center, north, and northeast of the country where 77 percent of the population resides, only 32 percent of renewable water is available. In the southeast where 23 percent of the population resides, 68 percent of the water was available in 2022 (14). Therefore, the resources are unequally distributed which obligates the State to develop alternative ways to bring water to places where there is a greater demand.

Mexico receives approximately 1,500,000 m3 of water in the form of precipitation annually; 71.7 percent evaporates and returns to the atmosphere (this percentage may fluctuate), 21.9 percent runs off through rivers and streams, and the remaining 6.3 percent infiltrates into the subsoil and recharges the aquifers (15).

Per capita availability in 2020 with a population of more than 126,000,000 was 3,663 m3. These numbers are of concern, especially considering the projections of the National Population Council which estimates that by 2050 the Mexican population will reach 150,800,000 inhabitants (14). This suggests that there will be an increased pressure on resources.

The renewable water availability per inhabitant in Mexico City is lower than in the rest of the country. In 2020, this indicator stood at just 70 m3 per year with a decreasing trend compared to 2019. Mexico's surface waters are made up of 51 rivers and streams through which 85.7 percent of the country's surface runoff flows. This is administered through 757 hydrological basins organized by 37 Hydrological Regions (14).

The division and administration of water helps to measure the average annual availability of national surface waters, which are published in the Official Gazette of the Federation to achieve integrated management, efficient use, and equitable distribution as well as to determine the granting of concessions.

The importance of groundwater lies in its function as a distribution network, filtration, reservoirs, and in the magnitude of the volume used by the main users. In Mexico, 39.4 percent of the total volume of water licensed for consumption (35,315 hm3 per year by 2020) comes from groundwater (14).

- From 2001 to 2020 the volume of surface water conces-sioned for consumption was 60.6 percent.
- In the same period, 39.4 percent was from groundwater sources (14).

Within these, the percentage of extraction was as follows:

- 75 percent agricultural
- 14.7 percent public supply
- 5.0 percent self-supplied industry
- 4.6 percent electric power excluding hydroelectricity (14).

The average availability of groundwater distributed across 653 aquifers is disclosed annually for the Hydrological Adminis-trative Regions. This measurement determines concessions and usage to avoid compromising the ecosystem balance. The situation of these aquifers is as follows:

- 32 have saline soils or brackish water.
- 18 have intrusion of seawater.
- 111 are overexploited (14).

The aquifer of the Mexico City Metropolitan Zone defined with the code 0901 by the National Water Commission (CONAGUA), is located in the central portion of the country. The surface area covered by the aquifer includes the 16 municipalities of the city.

According to data reported by the Public Registry of Water Rights (REPDA), the aquifer registered an annual extraction volume of 623.8 hm3 as of December 30, 2022 of which:

- 89.6 percent was potable water supply.
- 10.2 percent was for industrial use.
- 0.2 percent was for agricultural use.
- 0.0 percent was for domestic and livestock use (16).

For granting a concession it is necessary to know the availability of groundwater in order to calculate the possibility of exploiting the resource without endangering the ecosystem. For this purpose, CONAGUA carries out a calculation. The result indicates that for the aquifer of Mexico City there is no available volume to grant new concessions. On the contrary, there is a deficit of -469,629,914 m3 (16).

Regarding groundwater quality, the National Water Quality Measurement Network indicates that during the period from 2018 to 2020 this aquifer had good quality water in the recharge areas and lower quality water in the lowlands. However, it was within the limits established by the Mexican Official Standard (NOM) NOM-127-SSA1-2021. Iztapalapa municipality was an exception as iron and magnesium concentrations exceeded the maximum allowed for human consumption (16).

Surface waters contain various pollutants from industrial effluents, treatment plant discharges, sewage, soil lixiviates, mining industry waste, detergents, soaps, and run off from agricultural areas. Infections caused by these pollutions are cholera, dysentery, and other gastrointestinal disorders.

Aquifers on the other hand, are affected by the infiltration of urban wastewater and water from agricultural irrigation. This is worrying because there is pesticide content including chemical fertilizers (nitrates, phosphates), herbicides, insecticides, and fungicides. Additionally, leaks in the drainage networks introduce microorganisms and fecal coliforms.

In Mexico, water quality is measured by using biochemical oxygen demand (BOD5), chemical oxygen demand (COD), total suspended solids (TSS), and fecal coliforms (FC). According to these measurements in 2021, there was acceptable to excellent quality of water throughout the country (14).

However, since the year 2000, no updates have been made in the Mexican regulations regarding water quality that comply with WHO standards. It was not until May 2, 2022, that the update of the Official Mexican Standard NOM-127-SSA1-2021 was published in the Official Gazette of the Federation (17). This standard, which came into effect on April 27, 2023, establishes the standards for water intended for human consumption and the permissible quality limits. NOMs are mandatory technical regulations that ensure the quality, health, and harmonization of products and services for consumers. Their purpose is to promote and protect health.

The purpose of this analysis is not to conduct an exhaustive review of all WHO recommendations. Rather, the author focuses on highlighting, in general terms, the standards that Mexico incorporated in NOM-127 according to the Guidelines for Drinking-Water Quality as shown in Table 2.

TABLE 2
MICROBIOLOGICAL, CHEMICAL, AND RADIOLOGICAL POLLUTANTS AND
THEIR ACCEPTABILITY CRITERIA BASED ON NOM-127-SSA1-202117

MICROBIOLOGICAL	CHEMICAL	RADIOLOGICAL	ACCEPTABILITY
Regarding pathogens relevant to the management of drinking water supply systems, Mexico considers three of the 23 suggested by the WHO whose importance to health is considered high and whose persistence in the water supply is considered high or moderate. 1.Escherichia coli – Diarrheagenic 2.Cryptosporidium 3.Giardia lamblia	Of the eight chemicals of natural origin whose presence in water for human consumption can affect health as considered by the WHO, Mexico excludes barium and uranium. Even though arsenic is included, it is assigned a permissible limit of 0.025 mg/L versus 0.01 mg/L suggested by the WHO. Of the 21 chemical substances from industrial sources and inhabited areas that can affect health, Mexico omits dichloromethane, 1,4-dioxane and trichloroethylene. For cadmium, it considers a permissible limit of 0.005 mg/L versus 0.003 mg/L suggested by the WHO. For the 34 substances from agricultural activities, Mexico omits nitrate, nitrite, fenoprop, hydroxyatrazine, MCPAd, and methoxychlor. Of the 25 chemicals used in water treatment or from materials in contact with water, Mexico omits monochloramine, sodium dichloroisocyanurate, dibromoacetonitrile, dichloroacetonitrile, monochloroacetate, and n-nitrosodimethylamine. Of the pesticides contained in water that can affect health Mexico includes all of them as well as the permissible limit.	Mexico complies with the permissible limits for total alpha and total beta radioactivity in water omitting radon.	This characteristic addresses the taste, odor, and appearance of drinking water that may be unpleasant based on 30 parameters. Mexico does not include actinomycetes, invertebrate animals, zinc, chloramines, chlorobenzenes, chlorophenols, and hydrogen sulfide.

Source: Compiled by the author based on NOM-127-SSA1-2021 in comparison with the guidelines for drinking-water quality: fourth edition incorporating first addendum of the WHO (17)(6).
Table 2 shows the presence of permissible limits of microbio-

logical contaminants. Most of the pathogenic agents suggested by the WHO are not included in the national regulations.

Mexico established the same limits as the WHO for most of the chemical and radiological substances. Nickel and selenium are included. This is important because they are carcinogenic. Indicators such as *E. coli*, which are due to contamination of fecal origin, organisms such as *Giardia lamblia*, linked to intestinal disorders, and microcystin-LR, which can cause liver damage, are also included.

The omission of parameters that the NOM included prior to its reform is disconcerting. These parameters addressed physical and organoleptic characteristics such as color, odor, and taste, as well as turbidity with established permissible limits.

Compliance with NOM-127 is the responsibility of the Ministry of Health, the Federal Commission for Protection against Health Risks (COFEPRIS), and the state governments. This reform omits the inclusion of the CONAGUA in the surveillance which was the case in the 1994 NOM.

In this context, transition from regulations on paper to effective implementation in practice entails various challenges especially in the context of the human right to water. In Mexico, multiple agencies converge in the administration, regulation, control, and protection, as well as for monitoring and guaranteeing water quality.

In 2022, 61 percent of the Mexican population had safe drinking water supply services and in the case of Mexico City it was even higher with a percentage of 75.6. It should be noted that SDG 6 indicator measures those who have access to a reliable source of good quality drinking water at home (18).

In order to provide water to households, the Hydrological Regions in Mexico are fundamental both in their management and distribution since they carry out the analysis, diagnoses, and actions related to the quantity and quality of water as well as its exploitation and use (19). The Hydrological-Administrative Region XIII corresponds to Mexico City. It covers three states (Mexico, Hidalgo, and Tlaxcala) in addition to the 16 municipalities of the city. It has 62 water treatment plants, 120 municipal wastewater treatment plants, and 379 industrial wastewater treatment plants (19).

According to national statistics for this hydrological region, in 2022 the level of pressure on the water resources reached 128.6 percent. This level places the region in the red indicator which shows the maximum level of pressure on the system which uses 3,444.326 hm3 of renewable water annually. The annual concessioned volume is 4,428.856 hm3 (20). Although the indicator covers three states, the one that has the highest population is Mexico City.

In order for the Administrative Hydrological Region to provide water to the inhabitants of the city, water has to be extracted from the aquifer of the Valley of Mexico which represents 58 percent of the water supply, as well as from the import of the liquid through transfers from the Cutzamala and Lerma systems which together represent the remaining 42 percent (20). Recharge of this aquifer occurs through precipitation in elevated areas, a process that has had negative impacts during the last three years due to climate change. This situation has triggered droughts that have resulted in a decrease in dam storage with the Cutzamala System being one of the most affected. By February 2024, its useful water storage capacity stood at 38.7 percent (20).

Likewise, the Mexico City Water System (SACMEX), which is the authority that supplies and distributes drinking water and drainage services estimates that 12 percent of the water extracted to supply the city was deficient in its physicochemical characteristics where the main use for domestic consumption with 87 percent, 6 percent for mixed use, and 7 percent for non-domestic use (19).

It can, therefore, be concluded that surface water as well as groundwater in the city are severely overexploited which has caused a gradual sinking of the land and has damaged the supply and drainage system.

Bottling companies in Mexico and the Valley of Mexico aquifer

Currently, the world generates around 600,000,000,000 plastic bottles which are not recycled. This market for plastic bottles has increased by 73 percent, making it one of the fastest growing markets (21). It is estimated that 200 polyethylene terephthalate (PET) bottles are produced per year for every Mexican citizen (22).

These bottles are provided by a group of large companies

including Nestlé and Danone that introduced the use of PET in the 1980. A decade later, PepsiCo and Coca-Cola entered the business. The four large companies concentrated 27.5 percent of global production and marketing (23). By 2014, Primo Corporation was one of the largest companies in this market.

Market liberalization and deregulation in the 1980s and 1990s allowed foreign investment which resulted in a global concentration in few firms and increased sector dominance. Its expansion was possible due to institutional arrangements with the host state which allowed the use of public networks as well as springs and underground sources (23).

At the same time, the dismantling of the welfare state in the earlier decades led to a neglect of investment in public water systems resulting in a negative perception of the safety of tap water. Bottled water was favored as a safer option.

In the Latin American and Caribbean region, Mexico is the largest market for bottled water. The explanation is that earthquakes that occurred in 1985 affected the drinking water network and approximately 6,000,000 people were left without access to piped water mainly in the downtown area of Mexico City. The state provided water through water trucks and by delivery of water bags. An epidemiological surveillance system was set up to identify food-borne diseases. It showed that low socioeconomic areas were the most affected by diarrhea (24). Later in the 1990s, there was a cholera epidemic when the Mexican government promoted chlorination and boiling of tap water. Since then, consumption of bottled water has been favored with around 80 percent of the population using bottled water and 10 percent using water purified at home as their main source of drinking water (21).

The expansion of bottled water companies began to take place by developing partnerships. For example, Nestlé-Perrier bought local and regional companies. This also happened in the United States and Mexico to reduce costs and to take advantage of legislation and incentives (23). Likewise, the bottling company Danone with its presence in Mexico bought the national brand Bonafont by 1995. By 2001, it acquired the Pureza Aga brand, the second largest in Mexico (23).

Since then, there has been a culture of consuming bottled water and 20-liter jugs by the Mexican people despite the fact that filters can be bought in supermarkets. But these do not compare with the cost of bottled water. The average cost of one liter of drinking water for domestic use is two cents (0.0012 USD). Bottling companies sell this same volume for between 7.50 to 8.25 pesos (0.44 to 0.49 USD) (25).

In 2011, a survey was conducted on water consumption. All respondents indicated that they did not get quality water in the public supply which led them to consume bottled water. They were willing to pay a little more for good quality water. Bottled water is 500 to 1,000 times more expensive than tap water (23).

As Mexicans abandoned the consumption of tap water, water extraction concessions granted to beverage companies increased throughout the country. The production of one liter of bottled water involves the use of five to six liters of water.

And in the case of sugar-sweetened beverages, more than 10 liters of water (26).

The bottling companies that dominate the market in Mexico City are:

1. Local purification plants
2. Bonafont (Danone)
3. Electropura (PepsiCo)
4. E-pura (PepsiCo)
5. Ciel (Coca-Cola)

The most recent data for Mexico City indicates that in 2020:

- Per capita consumption of bottled water was 391 liters per person per day.
- 95.4 percent of household water consumption was in 20-liter jugs.
- Annual expenditure was 4,000,000,000 pesos for bottled water for households alone (27).

Bottling companies seek to ensure that their water is of the highest possible quality by using groundwater. Scientific and technological advances have promoted the exploration and exploitation of aquifers through drilling and pumping techniques which has simplified access to large volumes of water while concessions allow them an uninterrupted supply of water.

The Superior Audit Office of the Federation (ASF), which is the specialized technical body of the Chamber of Deputies, has technical and management autonomy and is in charge of auditing the use of federal public resources. An evaluation of the national water policy in 2019 indicated that CONAGUA did not have information that allowed for the analysis because there were not an enough number of macro-meters available to water operators to quantify the volume of water extracted by bottling companies and ensure that they did not exceed the maximum limits established in the titles granted to them. No information was available on the production costs incurred by water operators in the extraction, purification, and storage of water (28).

For the purpose of this analysis, the Public Registry of Water Rights (REPDA) of CONAGUA was consulted. It provides information on concessions, permits, or authorizations granted and specifies the holders. This consultation was carried out in April 2024 using filters requesting surface waters concessioned in the Aguas del Valle de México basin focusing the search only in Mexico City and in the industrial sector where bottling plants are concentrated. However, no results were obtained indicating that there are no concessions for surface waters in the general industrial sector.

The same exercise was conducted for groundwater where the REPDA indicated that 137 concessions were made from 1994 to 2023. When applying a first filter to identify the number of concessions granted during the current federal administration which began on December 1st, 2018, a total of 12 new concessions were detected of which four corresponded to bottling companies (29). Table 3 shows the volume of water extraction in cubic meters per year where out of the 137 records,

22 corresponded to bottling companies in Mexico City.

TABLE 3

ANNUAL VOLUME OF WATER EXTRACTED BY CORPORATE STAKEHOLDERS

TITLEHOLDER	CORPORATE PURPOSE	DATE OF REGISTRATION OF THE CONCESSION	VOLUME OF EXTRACTION OF NATIONAL WATERS (M3/YEAR)
EMBOTELLADOR A MEXICANA DE BEBIDAS REFRESCANTES	Production of soft drinks and purified water (Coca-Cola FEMSA)	07/10/2022	1,128,200.00
BEBIDAS PURIFICADAS, S. DE R. L. DE C.V.	Manufacture, distribution, and marketing of carbonated and non-carbonated beverages as well as bottle water (PepsiCo)	05/08/2022	511,392.00
ENVASADORAS DE AGUAS EN MEXICO, S. DE R. L. DE C.V.	Production and distribution of purified water Bonafont (Danone)	19/07/2022	332,400.00
EMBOTELLADORA METROPOLITANA, S. DE R. L. DE C.V.	Processing of soft drinks and other non-alcoholic beverages (PepsiCo)	08/06/2022	510,000.00
ENVASADORAS DE AGUAS EN MEXICO, S. DE R. L. DE C.V.	Manufacture and distribution of purified water Bonafont (Danone)	12/07/2017	160,526.00
EMBOTELLADORA AGA DE MEXICO, S.A. DE C.V.	Production, distribution and marketing of carbonated and non-carbonated beverages and purified water. (AGA-Danone)	14/05/2009	20,514.00
BEBIDAS PURIFICADAS, S. DE R. L. DE C.V.	Processing of soft drinks and other non-alcoholic beverages (PepsiCo)	28/07/2008	206,759.00
EXTRACTOS Y MALTAS SA DE CV	Manufacture, purchase, sale, import, and export of extracts and malts (Heineken)	23/07/2004	100,000.00

EXTRACTOS Y MALTAS, S. A. DE C. V.	Brewing production (Heineken)	28/07/1995	218,832.00
CERVECERÍA MODELO, S.A. DE C.V.	Brewing production (Anheuser-Busch InBev)	18/04/1995	95,562.00
PROPIMEX, S. A. DE C. V.	Production and distribution of soft drinks (Coca-Cola FEMSA)	22/12/1994	486,880.00
PROPIMEX, S. A. DE C. V.	Production and distribution of soft drinks (Coca-Cola FEMSA)	22/12/1994	110,000.00
PROPIMEX, S. A. DE C. V.	Production and distribution of soft drinks (Coca-Cola FEMSA)	11/11/1994	817,255.00
Total			14,371,293.00

Source: *Author's compilation based on information from the Public Registry of Water Rights of Mexico (29).*

The water extractive industry described here can dispose of this resource through indirect concessionaires and intermediaries that provide their services using different names. It is a challenge to identify them and to know the relationship with transnational companies shown in Table 3. At least three of the large bottling companies produce, distribute, and bottled water and carbonated beverages. It is worth mentioning that the corporate purpose had to be researched in open sources since REPDA does not provide such information.

126

In addition, it was decided to include those concessions granted to brewing companies which have significant water hoarding throughout the country considering that for each liter of beer 3.63 liters of direct water are required (30).

The grand total reveals an annual hoarding of more than 14,000,000 cubic meters of water. Considering also that according to Article 24 of the National Waters Law in force, the term of the concession or assignment for the use or exploitation of national waters ranges from 5 to 30 years and can be extended. The REPDA does not disclose whether it is a first concession or a renewal (31).

Although Mexico City has reported a level of safely managed drinking water supply for 75.6 percent of its population, the aquifer of the Metropolitan Zone of Mexico City indicated an unsustainable level of overexploitation as there is no available volume to grant new concessions. On the contrary, it shows a deficit. Also, the availability of renewable water per inhabitant in Mexico City is one of the lowest in comparison with the rest of the country, standing at barely 70 m3.

This stress has repercussions on the quantity and quality of fresh water which restricts the optimal fulfillment of the Human Right to Water. Specifically, availability in relation to demand and the lack of adequate infrastructure to access water has forced affected users to allocate a portion of their salary for the purchase of water for daily activities and for human consumption through trucks, with pipes, and bottled water.

The sale of bottled water has become a multi-million-dollar

industry and has been incorporated into the daily routine of millions of people. The growing influence of Transnational Corporations (Tncs) that operate beyond national regulations combined with the loss of national sovereignty and advertising, the perception of water safety, and consumption habits have allowed this industry, which was originally conceived as a public service, to flourish and transform into a lucrative business.

Health impacts of water quality on Mexicans

Anthropogenic overexploitation of the aquifer of the Valley of Mexico, neglect of infrastructure, and low availability due to droughts, recently experienced create an optimal scenario for the deterioration of water quality.

These problems have led to gastro-intestinal infections. There have also been problems due to the presence of chemical compounds of industrial origin in the water. There is limited official information and the only indicator that was found in this regard was updated in 2018. According to the Ministry of Health, the morbidity rate attributable to waterborne diseases results in thousands of cases per 100,000 inhabitants has shown a declining trend since 2000. It has decreased from an annual rate of 7.5 to 4.8 in 2018 (32).

For this parameter, only intestinal amoebiasis, ascariasis, cholera, dengue fever (non-severe and severe), mild malnutrition, moderate malnutrition, severe malnutrition, scabies, yellow fever, paratyphoid A, typhoid fever, viral hepatitis A, leptospirosis, onchocerciasis, and other intestinal infections due to protozoa, other salmonellosis and malaria were considered (32).

Most diseases are multifactorial which means that they are caused by various agents and are due to exposure and/or consumption of polluted water. The drinking water service is provided by the municipalities that oversee the enforcement of NOM-127 and guarantee water safety.

While the public water supply is perceived to be of poor quality, bottled water is also not safe. In a study conducted between 2016 and 2017 to determine the microbiological quality of bottled water from purification plants located in different municipalities of Mexico City, 111 samples of 20-liter jugs were analyzed. Samples analyzed for bacteriological quality showed that 72.9 percent did not meet the standards established by the NOM-041-SSA1-1993 which establishes the sanitary specifica-tions for bottled water (33).

Additionally, establishments that refill water bottles under the supposed purification service have increased over time and a considerable number of them are not regulated. They continue to operate without major repercussions. Although the National Statistical Directory of Economic Units records that in Mexico City there are 2,160 establishments, it is estimated that there was an increase of more than 50 percent of these irregular businesses with around 24,000 establishments by 2020 throughout the country (34)(35).

As there are few options available to citizens, there is no

guarantee of the Human Right to Water. Even the Human Rights Commission of Mexico City states that there were 1,537 complaints related to the human right of water and sanitation during the period 2012-2019 which averages an annual number of 192 complaints (36).

The top six reasons that prompted complaints were:

1. Unjustified interruption of service
2. Excessive or unjustified charges
3. Omission of the authority to bring the service closer to the communities, and educational and work centers.
4. Omission or delay on the part of the authority to repair water leaks.
5. Location of public water service in unhealthy conditions.
6. Omission or delay on the part of the authorities to report, in a timely manner, the suspension of the service (36).

Information available on water and sanitation and their relationship with health is inadequate and lacks coordination which limits our understanding of the problem. Nevertheless, the social reality described by citizens and portrayed in the news points to an emergency situation.

As mentioned, Mexico City is divided into 16 municipalities within which there is a notable variability in terms of water quality, quantity, and availability. During the first four months of the year 2024, complaints and demands addressed to the Federal Government were in relation to the following situations:

- Inhabitants of the Benito Juarez municipality reported contamination from oils and lubricants that resulted in water having with a strange odor, color, and taste. They reported problems such as dermatitis, infections, stomach pains, and burning eyes (37).
- Inhabitants of the Tláhuac municipality reported receiving water with a yellow tone (38).
- Inhabitants of the Álvaro Obregón municipality reported that there was a smell of hydrocarbons because of which the Alfonso XIII Water Well had to be guarded (39).
- Inhabitants of the Tlalpan municipality reported lack of service and cuts in supply (40). The Tlalpan district reported water shortages and leaks that had been present for five months (41).
- Inhabitants of the Magdalena Contreras municipality pointed out that the quality of the water they receive had been affected by the construction of mega housing projects. It had a yellowish tone and a metallic odor (40).
- The inhabitants of the Coyoacán municipality experienced a lack of water supply. They purchased water through private suppliers, paying prices ranging from 1,600 pesos (35.3 USD) for a 10,000-liter pipe to 3,500 pesos (205.9 USD) for a 20,000 litre pipe (42).
- Inhabitants of the Miguel Hidalgo municipality asked the Mexico City Government to send drinking water pipes to address the shortages they were experiencing (43).
- Workers at La Viga Market in Iztapalapa suffered water shortages and had to pay from 70 pesos (4.1 USD) per drum to 1,800 (105.9 USD) for the pipeline service, buying up to three trucks of water per week (44). The worst quality of water with unpleasant odor, color and taste was present in

this municipality of Mexico City.

· The inhabitants of the Iztacalco municipality reported a lack of water supply for the last seven years. The water that they received had a turbid color (45).

A review of the national press points to a problem that is repeated in several municipalities of Mexico City. People complained of shortages and leaks that were unattended for months. They also complained that there was a charge for services that were not provided and that the water was of poor quality. They mentioned that there was privatization and commercialization of the liquid through concessions and support for companies not only in the beverage and brewing industries, but also in the automotive, real estate, and agro-industries sectors among others.

Concluding comments

Mexico City faces several challenges. Negligence in the maintenance of water infrastructure has led to low water quality and inequity in water supply as well as shortages due to leaks or droughts. There is also a pattern of over-concession of water by the agro-industry which privileges its allocation by legal means thus promoting its commoditization. At the same time, there is a lack of coordination among the authorities responsible for water monitoring and management which complicates the implementation of effective measures to address water quality and supply problems.

Water quality in Mexico City does not reflect the reform of NOM-127-SSA1-2021. Although this did not include all the standards of the WHO Guide, it did include a significant number of chemicals produced by the agro-industry that had not been properly regulated for more than 22 years. It is considered an advance by the State in the achievement of the Human Right to Water in compliance with the principle of progressiveness.

While citizens must adopt a rational approach, it is essential to control those who hoard water and make intensive use of it. To guarantee fair taxation regulation, legal compliance and control of unfair practices, consumer protection under the lens of human rights and anti-corruption measures is needed so that there is transparency and accountable regarding the actual volume of water withdrawal.

It is difficult for citizens to use water rationally if the agro-industry continues to expand its water footprint without effective regulation. It is also necessary to ensure equity in supply. Complaints to the municipalities indicate that water is deviated and prioritized for areas of higher economic level leaving behind the most vulnerable sector of the population who often pays more for water.

During the recent water shortage in Mexico City, the civil organization known as Agua Capital presented an initiative to address this problem. Its proposal included the regulation of aquifers, artificial recharge techniques, cancellation of irregular use, integrated watershed management, and increased wastewater treatment (46). This proposal did not mention guarantee of the human right to water. Nor did it place people at the center. This proposal emanated from an association formed by Coca-Cola, Grupo Modelo, Grupo FEMSA and Cervecería Cuauhtémoc, among others who hold 83 percent of the concessions for water use.

In addition to the high concentration of water, the production of its products generates resins that are used to produce rigid HDPE plastic and PET which determines the market. During the production process, these resins release toxic chemicals into the atmosphere and into the bodies of water. These are ethylene oxide, benzene, and xylenes which are known to cause cancer, and which can damage the nervous system (23).

Water in Mexico and the world is in dispute and this scenario is reproduced in developing countries where the expansion of capital increasingly needs other territories in the name of development and progress, but for whom? Environmental change does not occur in a political vacuum. It is intrinsically linked to social, material, and symbolic disputes. Citizens must, therefore, question and demand from those who have the power to decide on the use, control, appropriation, and transformation of water.

Municipalities lack budgets and operational capacity. According to the National Association of Water and Sanitation Entities of Mexico (ANEAS), service providers collect 68,000,000,000 pesos annually in fees which is barely enough to cover their current expenses. While tariffs are an issue that must be addressed, they should not fall into the logic of profit maximization and exclusion of vulnerable groups. Water should remain affordable, and prices should reflect its actual usage to consumers.

Likewise, it is estimated that approximately 90,000,000,000 pesos are required to rehabilitate the 12,000 kilometers of piping that make up the service network. This is equivalent to the budget of 17 and a half years of the SACMEX or equivalent to two years national sales of bottled water (47).

As the bottled water market expands, it becomes increasingly important to strengthen and reform national legislation. Mexico has a long-standing debt in this area since 2012 when the Human Right to Water and Sanitation was elevated to constitutional rank through the reform of the sixth paragraph of Article 4 of the Constitution which resulted in the obligation of the Legislative Power to formulate a General Water Law that guarantees the protection of this right within 360 days.

Realization of this right constitutes a demand for social justice and compliance with responsibilities acquired from the signing of international treaties and their adaptation within national regulations where priority in public supply should continue to dominate over any other use. These responsibilities are outlined in the General Comment 15 of the International Covenant on Economic, Social and Cultural Rights, of which Mexico is already a part.

Continuing and reaffirming the human right to water implies using this tool as a direct challenge to the power relations and inequalities rooted in the discourse of capitalist development *ad infinitum* in order to place the human person and his or her dignity at the center. Recalling the inherent characteristics of human rights which are universal, absolute, inalienable, and non-seizable, this last characteristic stands out for its

135

indispensable character for subsistence.

References

1. Rockstrom J, Steffen W, Noone K, Persson A, Chapin FS III, Lambin E, et al. Planetary boundaries: Exploring the safe operating space for humanity. Ecology and Society. 2009. 14(2): 32. http://dx.doi.org/10.5751/es-03180-140232
2. UNESCO. The United Nations world water development report 2023: Partnerships and cooperation for water. 2023. Paris: Government of Italy and Regione Umbria: 2023. https://unesdoc.unesco.org/ark:/48223/pf0000386807
3. ESC. General comment no. 15 Geneva. United Nations: Economic and Social Council. 2002 Nov 11-29. https://www2.ohchr.org/english/issues/water/docs/CESCR_GC_15.pdf
4. UN. Resolución aprobada por la Asamblea General el 28 de julio de 2010. United Nations. 2010. https://digitallibrary.un.org/record/687002/files/A_RES_64_292-ES.pdf Spanish.
5. UN. Instantánea del ODS 6 en México. UN Water. https://www.sdg6data.org/es/country-or-area/Mexico. Spanish.
6. WHO. Guidelines for drinking-water quality: fourth edition incorporating the first and second addenda. 4th ed. London, England. IWA Publishing. 2011; 583. https://iris.who.int/bitstream/handle/10665/352532/9789240045064-eng.pdf?sequence=1
7. Ortiz PE. Contaminación Microbiológica del Agua para consumo humano. Honduras. Consejo Nacional de Agua potable y Saneamiento; 2023. https://conasa.hn/files/64/Capacitaciones-2/414/Webinario-3-Contaminacion-

Microbiologica-del-agua-PEO2023.pdf Spanish.

8. WHO, UNICEF & World Bank. State of the world's drinking water: an urgent call to action to accelerate progress on ensuring safe drinking water for all. Geneva: World Health Organization. 2022 https://iris.who.int/bitstream/handl e/10665/363704/9789240060807-eng.pdf?sequence=1

9. Communicable Disease Center. Chemicals that can contaminate tap water. Cdc.gov. 2022 https://www.cdc.gov/ drinking-water/causes/chemicals-that-can-contamina te-tap-water.html?CDC_AAref_Val=https://www.cdc.g ov/healthywater/drinking/contamination/chemicals.ht ml

10. WHO. Arsenico. Who.int. 2022. https://www.who.int/es/ news-room/fact-sheets/detail/arsenic Spanish.

11. Red del Agua UNAM. Contaminantes emergentes en el agua: causas y efectos. Impluvium. 2021 Oct -Dec; 17. https://www.agua.unam.mx/assets/pdfs/impluvium/nu mero17.pdf Spanish.

12. Bruce J. Lesikar, Michael F. Hare, Janie Hopkins, editor. Drinking water problems: Radionuclides. U.S.A: Texas Water Resources Institute of Texas. 2021. https://twon.ta mu.edu/wp-content/uploads/sites/3/2021/06/drinking- water-problems-radionuclides.pdf

13. Vázquez del Mercado Arribas R. Huella hídrica en México: Análisis y perspectivas. 1st ed. Mexico: IMTA; 2017. http://repositorio.imta.mx/handle/20.500.12013/ 1714 Spanish.

14. SEMARNAT, CONAGUA. Numeragua 2024. 1st ed. México: SEMARNAT. 2022. https://iwlearn.net/iw-projects/orga nizations/920. Spanish.

15. CONAGUA. Estadísticas del Agua en México 2021. Ciudad

de México: SEMARNAT. 2022 Oct. https://files.CONAGUA
.gob.mx/CONAGUA/publicaciones/Publicaciones/EAM%
202021.pdf Spanish.

16. CONAGUA. Actualización de la disponibilidad media an-
ual de agua en el acuífero de la zona metropolitana de
la Ciudad de México (0901). Ciudad de México. 2024.
https://sigagis.conagua.gob.mx/gas1/Edos_Acuiferos_
18/cmdx/DR_0901.pdf Spanish.

17. DOF. NORMA Oficial Mexicana NOM-127-SSA1-2021.
Agua para uso y consumo humano. Límites permisibles
de la calidad del agua. Gob.mx. https://www.dof.gob.mx/
nota_detalle.php?codigo=5650705&fecha=02/05/2022.
Spanish.

18. Secretaría de Economía, INEGI, 6. Agua limpia y
saneamiento. Agenda 2030. https://agenda2030.mx/
ODSind.html?ind=ODS006000050030&cveind=618&
cveCob=99&lang=es#/Metadata Spanish.

19. CONAGUA. Gobierno Ciudad de Mexico, SACMEX. La
Cuenca del Valle de México Guía para el maestro. Ciudad
de México: SACMEX. 2020. https://aplicaciones.sacmex
.cdmx.gob.mx/libreria/biblioteca/libros/2021/Gui%C3%
ACa%20libro%20agua_SACMEX_PRINTok_logoverde.
pdf Spanish.

20. CONAGUA. Sistema nacional de información del agua.
Gob.mx: CONAGUAhttps://sinav30.CONAGUA.gob.mx:8
080/SINA/?opcion=gpresion. Spanish.

21. Bouhlel, Z., Kopke, J., Mina, M. & Smakhtin, V. Global
bottled water industry: A review of impacts and trends.
Canada: United Nations, University Institute for Water,
Environment and Health. 2023. http://collections.unu.ed
u/view/UNU:9106

22. Comisión Nacional de Áreas Naturales Protegidas. Con-su-mismo plástico. gob.mx: Gobierno de Mexico. 2018 Nov 23. https://www.gob.mx/conanp/es/articulos/con-su-mismo-plastico?idiom=es. Spanish.

23. Montero D. Transnacionales, gobierno corporativo y agua embotellada. 1st ed. Ciudad de México: Universidad Autónoma Metropolitana; 2015. https://www.researchga te.net/publication/313283977_Montero_Contreras_De lia_2015_Transnacionales_gobierno_corporativo_y_ agua_embotellada_El_negocio_del_siglo_XXI_Mexi co_Universidad_Autonoma_MetropolitanaEdiciones_ del_Lirio Spanish.

24. Ruiz Matus C, Cárdenas Ayala V, Koopman J, Herrera Bas-tos E, Montesano Castellanos R & Hinojosa M. Enfermedad diarreica después de los sismos de 1985 en México. Salud Pública.1987; 29 (5): 399-405. https://pubmed.ncbi.nlm. nih.gov/3424005/ Spanish.

25. Villanueva D. México, el mayor consumidor de agua em-botellada en el mundo. Jornada La Jornada. (Mexico Ed.) 2021 April 02. https://www.jornada.com.mx/notas/2021/ 04/02/economia/mexico-el-mayor-consumidor-de-ag ua-embotellada-en-el-mundo/. Spanish.

26. Sistema de Aguas de Huixquilucan. El impacto del agua embotellada en México. Mexico. SAH. 2023 May 16. https://sah.gob.mx/blog/2023/05/16/el-impacto-del-a gua-embotellada-en-mexico/. Spanish.

27. Montero D. El agua embotellada y el Covid-19. Unam.mx. Mexico. UNAM. 2020 May 27. http://www.agua.unam.mx /covid19/assets/pdf/AguaEmbotallada_DeliaMontero.p df. Spanish.

28. ASF. Evaluación número 1371-DS: Evaluación de la política

hídrica nacional. Mexico ASF. 2019. https://www.asf.go
b.mx/Trans/Informes/IR2019c/Documentos/Auditorias/
2019_1371_a.pdf. Spanish.

29. Comisión Nacional del Agua. Consulta a la base de datos
del REPDA. Mexico: Gob.mx. 2002. https://app.CONAGU
A.gob.mx/consultarepda.aspx. Spanish.

30. Delgado Ramos GC. Apropiación de agua, medio ambi-
ente y obesidad: los impactos del negocio de bebidas
embotelladas en México. 1st ed. México: UNAM-CEIICH;
2014. https://ru.ceiich.unam.mx/handle/123456789/284
8 Spanish.

31. Cámara De Diputados del H. Congreso de la Unión. Ley De
Aguas Nacionales. México: DOF. https://www.diputados.
gob.mx/LeyesBiblio/pdf/LAN.pdf. Spanish.

32. Secretaría de Salud. Anuarios de Morbilidad 1984-2018
Distribución de casos nuevos de enfermedad por fuente
de notificación. Mexico: Secretaria Salud. 2020 Feb.
https://apps1.semarnat.gob.mx:8443/dgeia/compartido
s/pdf/COM_MORB_HID.pdf. Spanish.

33. Cerna-Cortes JF, Cortes-Cueto AL, Villegas-Martínez D,
Leon-Montes N, Salas-Rangel LP, Rivera-Gutierrez S, et
al. Bacteriological quality of bottled water obtained from
Mexico City small water purification plants: Incidence and
identification of potentially pathogenic nontuberculous
mycobacteria species. ELSEVIER. 2019 Oct 02; 306:1-6.
http://dx.doi.org/10.1016/j.ijfoodmicro.2019.108260.

34. Instituto Nacional de Estadística y Geografía (INEGI).
Directorio Nacional de Unidades Económicas DENUE. Mexico: INEGI. 2014. https://www.inegi.org.mx/app/mapa/
denue/default.aspx. Spanish.

35. Caballero SCS. Coordinación de Comunicación Social: De-

nuncian incremento descontrolado de establecimientos purificadores de agua. Mexico: Gob.mx. 2021 Sep 12. https://comunicacionsocial.senado.gob.mx/informacio n/comunicados/626-denuncian-incremento-descontro lado-de-establecimientos-purificadores-de-agua Span- ish.

36. Osorno Córdova C. Situación del derecho al agua en la Ciudad de México en tiempos de COVID-19. Métodhos. 2020 Jun 30;1(18):34-55. https://revista-metodhos.cdhc m.org.mx/index.php/metodhos/article/view/135/2020_ 18_metodhos_articulo_2_html. Spanish.

37. Camhaji E. Agua contaminada en Ciudad de México: 10 respuestas y recomendaciones de un especialista EL PAÍS (Mexico Ed). 2024 April 11. https://elpais.com/mexico/ 2024-04-12/agua-contaminada-en-ciudad-de-mexico -10-respuestas-y-recomendaciones-de-un-especialist a.html. Spanish.

38. Online La Razón. Denuncia el PAN "agua amarillenta" ahora en Tláhuac. La Razon (Mexico ed). 2024 April 16. https://www.razon.com.mx/ciudad/denuncia-pan-agua -amarillenta-tlahuac-573165. Spanish.

39. Sosa Santiago MF. Cierran pozo de agua contaminada en la alcaldía Álvaro Obregón que abastecía a Benito Juárez. El Economista (Mexico ed). 2024 April 10. https://www.el economista.com.mx/estados/Cierran-pozo-de-agua-co ntaminada-en-la-alcaldia-Alvaro-Obregon-que-abast ecia-a-Benito-Juarez---20240410-0049.html Spanish.

40. Uribe B. Preocupa sobreexplotación de pozos en Tlalpan. REFORMA (Mexico ed). 2024 March 10. https://www.r eforma.com/preocupa-sobreexplotacion-de-pozos-en- tlalpan/ar2770721. Spanish.

41. León A. Padecen 94 colonias en Tlalpan por escasez de agua. REFORMA (Mexico ed). 2024 March 25: https://www.reforma.com/padecen-94-colonias-en-tlalpan-por-escasez-de-agua/ar2779023. Spanish.

42. Sánchez Bolaños Á. Denuncian falta de agua en Coyoacán; afecta a 380 familias. La Jornada (Mexico ed). 2024 Jan 26. https://www.jornada.com.mx/noticia/2024/01/26/capital/denuncian-falta-de-agua-en-coyoacan-que-afecta-a-380-familias-879. Spanish.

43. López J. Se incrementan solicitudes de pipa en Polanco y Lomas de Chapultepec. Excelsior (Mexico ed). 2024 March 04. https://www.excelsior.com.mx/comunidad/se-incrementan-solicitudes-de-pipa-en-polanco-y-lomas/1639160. Spanish.

44. Vargas A & Mora K. Comerciantes de La Nueva Viga buscan agua para mantener fresco su producto. El Sol de México (Mexico ed). 2024 Feb 13. https://www.elsoldemexico.com.mx/metropoli/cdmx/comerciantes-de-la-nueva-viga-buscan-agua-para-mantener-fresco-su-producto-11433926.html. Spanish.

45. Bravo EM. Vecinos de la colonia Agrícola Oriental siguen sin agua potable. La Jornada (Mexico ed). 2024 March 14: Sect. Capital. https://www.jornada.com.mx/2024/03/14/capital/032n1cap. Spanish.

46. Montalvo O. Encubre UNAM a empresas sobreexplotadoras de agua. Diario Basta! (Mexico ed). 2024 March 11: Sect. CDMX https://diariobasta.com/2024/03/11/encubre-unam-a-empresas-sobreexplotadoras-de-agua/ Spanish.

47. Dirección General de Comunicación Social. Hogares capitalinos gastan 4 mil millones de pesos en agua embotel-

lada al año. Gaceta UNAM. (Mexico ed). 2019 Jul 07 https://www.gaceta.unam.mx/hogares-capitalinos-gas tan-4-mil-millones-de-pesos-en-agua-embotellada-al-ano/ Spanish.

6

Human Behavior in Urban Ecology: Mental Health and Climate Change

Abstract

This contribution is presented through the lens of human interaction with the built environment using an interdisciplinary approach of psychology, environmental science, and architecture
/ urban design. The chapter introduces the "human eco-centric
design" concept advocating for a built environment design
that integrates human behavior/interventions, climate, and
health. The study provides an overview of: (i) the relationship
between the built environment, climate, and health; (ii) mental
health in built environments, and the position on diversity and
inclusion highlighted by focusing on vulnerable populations,
(iii) design recommendations for the building stocks through
the perspective of human eco-centric design; and (iv) potential
areas for future research and design in the building sector to
combat the climate crisis and improve mental health.

Keywords: Built environment, climate crisis, vulnerable population, mental health, human eco-centric design

Authors

Zahida Khan, Assistant Professor of Architecture, Ball State University, Muncie- IN [zahida.khan@bsu.edu]

Piyush Khairnar, Assistant Professor of Architecture, Texas Tech University, Lubbock-TX [pikhairn@ttu.edu]

Fatima Khan, Student Researcher, University of Illinois in Chicago, Chicago-IL [fkhan89@uic.edu]

Introduction – Human eco-centric design

Cities are channels of social, cultural, environmental, and economic transactions, where people constantly interact with the built environment. The constant rise in world population is leading to a roar in urbanization. The 2018 urban population, estimated at 55.3 percent of the total world population, is expected to reach 68 percent by 2050 as per the World Urbanization Prospects Report (1). With the increasing trend in urbanization, it has become very important to implement the 2030 Agenda for Sustainable Development Goals (SDGs), especially SDG-11 "to make cities and human settlements inclusive, safe, resilient, and sustainable" (2). This agenda requires a human-centric approach to designing built environments using people as the key performance metric- a bottom-to-top approach for design. Indoor-outdoor spaces such as a room in a building or an outdoor plaza in an urban block inhabit people and facilitate human interaction with the built environment.

Moreover, human behavior resulting from this interaction provides critical insights into assessing people's mental and physical health and building performance. Studies show that such interactions involve an interrelationship between built environments and humans aiming to improve the quality of life in sustainable cities and communities (3). In this age where we are constantly striving to address two of the prime global issues, namely climate crisis and mental health, the topic of human-building interaction becomes vital. It requires introspective discussion as far as the current built environment trends are concerned. To understand climate and health within the premise of the human–building interaction, we must first understand their relationship with buildings.

Beginning with health historically, we have seen that health has always been one of the significant drivers of built environment design. Building codes in their early years focused on the safety of people. Later, physical health and comfort started being incorporated into the legislatures and standards that drove the design of the built environment. For example, ASHRAE standards 62 focus on thresholds for ventilation rates required for indoor air quality acceptable to human occupants (4). Recently there has been increasing interest in identifying the connection between buildings and mental health (5)(6). WHO identifies one in every eight people globally is facing a mental disorder or health condition (7). These conditions are more prominent in vulnerable populations with growing or degenerating mental development such as children or older adults. Designing for such diverse populations is often limitedly considered in the building design and construction processes due to their specific needs. Although initiatives such as Universal Design (8) and the WELL building standard outline the design guidelines to promote health in general, a specific focus on mental health has yet to be addressed, especially for vulnerable populations with specific needs. This aspect would be understood by studying the behavioral concept in psychological science. And in a specific dimension of architectural spaces, it would be best to explore human-building interaction in these populations. While we are talking about the research gap in the area of architectural spaces and mental health of vulnerable populations, its relationship to the climate crisis remains a blind spot that requires immediate attention. To explore this, it makes sense to first understand the relationship between built environments and the climate crisis.

Buildings involve activities such as energy consumption, construction waste, building materials, landscape, space planning, building systems, etc. which, if not addressed appropriately, can have a detrimental impact on the climate. The global energy demand by buildings in 2019 was equivalented to 31 percent of the global final energy demand, as per the Intergovernmental Panel on Climate Change (IPCC) 2022 report (9). This increased energy consumption along with the greenhouse gas (GHG) emissions in the building sector are contributing to the climate emergency. Architects and engineers in the building industry are making rigorous efforts to mitigate the effects of climate change and engage in high-performance built environments. Innovative design strategies are fostered to create architectural spaces (indoor and outdoor) aiming to ensure occupant comfort and reduce global warming. Besides design strategies, systems, and space planning, recent studies show that the occupants and their behaviors have a significant impact on energy consumption. This calls for measures to consider occupant behavior in high-performance buildings. Moreover, occupant behavior is a resourceful metric for measuring the impact of indoor and outdoor design on occupant health. In this age of technology, people spend most of their time indoors leading a sedentary lifestyle. This has increased morbidity rates resulting from a lack of physical activity and poor indoor environmental conditions. Hence, designing good indoor and outdoor spaces is not just a concern of the climate crisis, but also of health especially mental health. Such built environments incorporating human behavior for uplifting public health and mitigating the climate crisis are called "human eco-centric design".

Figure 1
Human eco-centric design

Human eco-centric design

Source: *Created by author (Zahida Khan)*

The interrelationship between built environments, mental health, and climate emergencies can be established by using human activities and behaviors as the epicenter of human eco-centric design. The complex topic of human interventions requires an interdisciplinary approach, primarily through health science, environmental science, and building science. The following sections dive deeper into the relationship between built environments, climate, and mental health.

Urban ecology and climate crisis

Anthropocene and planetary boundaries

The combination of specific environmental conditions has allowed our blue planet to sustain life, subject to the delicate balance of geological and atmospheric conditions. In the Neolithic period, the invention of agriculture was followed by the beginning of civilization. In a relatively short period after that, humans metamorphosed from living harmoniously with nature to being a dominant force in shaping our planet. Anthropocene marks a new ecological epoch where human activity singularly drives environmental change that needs attention (10). Anthropocene is defined by significant alterations ranging from deforestation and urbanization to pollution and climate crisis caused by human activities on the planet.

In 2009, Johan Rockström and Will Steffen introduced the limits to the impact of human activity on the planetary systems. These planetary boundaries describe multiple limits that range from biosphere integrity and climate change to land use change, biogeochemical flows, and novel entities. These limits represent the threshold beyond which our environment may not be able to self-regulate. These irreversible environmental changes threaten the stability and resiliency of our planet (11). The rate of reaching or crossing these planetary boundaries has accelerated in the post-industrial revolution world. Greenhouse gas emissions are rising with the increasing trend of burning fossil fuels, industrial agriculture, and rapid deforestation due to unprecedented urbanization, triggering global warming. The negative impacts of these anthropogenic activities drastically affects the weather patterns, sea levels, and ecosystems around the globe (12).

Buildings, carbon footprint, and the climate crisis

The anthropogenic act of building is a response to unfavorable environmental conditions and the need to achieve comfortable living conditions. From the Primitive Hut, theorized by Marc-Antoine Laugier as the fundamental basis for creating architecture to the most recently competed indoor ski resort in the middle of a desert in Dubai, the practice of architecture has evolved significantly. Buildings and surrounding areas are vital for human habitation as they provide shelter, workplace, and places for recreation and social interaction. These buildings and their surroundings collectively form our neighborhoods, towns, and cities, occupying large areas of land and extracting significant natural resources directly and indirectly.

For example, a typical single-family dwelling in North America requires building materials such as timber, stone, metal, insulation, and several plastic products during its construction. The same house requires potable water, electricity, and natural gas when it is occupied. Ultimately, it requires demolition and transportation of building waste to the landfill or recycling facility. One can now start to think about the potential monetary investment along with the amount of natural resources required for this house in its entire life span. If we expand this concept, the three distinct stages in the life of every building can be characterized as: (i). Construction phase, (ii). Operation phase, and (iii). End of life. It is important to note that the resources spent extracting raw materials that are processed to achieve building materials are also energy-intensive. Infact, all the products, such as electronic appliances, food related products, and building fixtures also contribute to the extraction of natural resources and the energy footprint during the "Operational phase" of the building. Understanding the energy consumption of buildings and their impact on the natural environment is complex and requires a sophisticated and comprehensive approach.

Figure 2
Life cycle stages for buildings and associated global carbon emission and energy consumption as per UNEP 2024 report

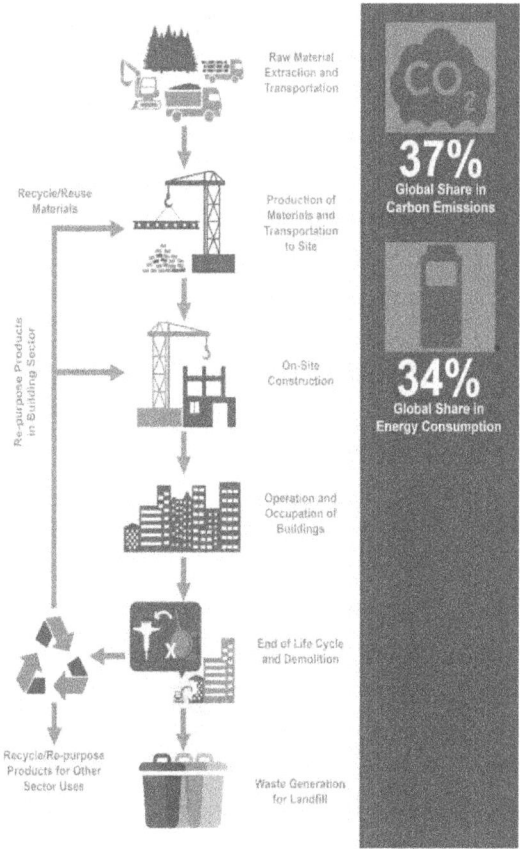

Source: *Created by author (Piyush Khairnar)*

The impact of these anthropogenic activities on natural systems can be quantified with two metrics: energy footprint and carbon emissions. The former measures energy consumption during the life of a product, building, organization, or territory, while the latter measures the amount of greenhouse gas emissions generated. It is crucial to track carbon emissions to demonstrate the impacts of the energy footprint in terms of greenhouse gas emissions, the most prominent cause of global warming. The building and construction sector contributes significantly to global energy demand and carbon emissions. In 2022, the share of the building sector in the total global energy demand was 34 percent, whereas the carbon emissions generated by the production and operation of buildings globally were recorded at 37 percent (13). It is important to note that the demand for energy to operate buildings amounts to 30 percent and the remaining 4 percent comes from the energy required to produce the construction materials. In contrast, carbon emissions for the operational phase of buildings account for 27 percent of the global share and material manufacturing and fabrication generate 10 percent of the global carbon emissions (13).

The increasing trend in carbon emissions signifies a rising quantity of greenhouse gases in the troposphere, leading to global warming. The planet's temperature increase can cause catastrophic failures of the earth's systems, leading to climate change, as witnessed by the environmental disasters in recent decades. These climatic disturbances are causing unprecedented draughts in regions across the globe while others are faced with catastrophic rainfall, leading to the destruction of the ecological balance. Global sea levels have also increased significantly in the last century. According to a study led by the United Nations, the rate of sea level rise doubled from 2.27 mm (0.089 inch) in 1993-2002 to 4.62 mm (0.18 inch) in 2013-2022 (14). This rise in sea level is triggered by increased global temperature, causing ice loss on land, melting glaciers and ice sheets in polar regions, and thermal expansion.

Globally, an estimated 410 million people are at risk from coastal flooding, resulting in mass global migration (15).

The relationship between buildings, carbon emissions, and the climate crisis is complex. The built environment is not only a significant driver of global climate change in the post-industrial world, but it also suffers from the devastating impacts of the climate crisis. A systematic and comprehensive approach is needed to explore the workings of the built environment and its effects on the climate. The pressing challenges of the climate crisis demand an urgent response to reduce global carbon emissions and energy consumption.

Human thermal comfort, built environment, and energy consumption

Exploring human psychological and physiological interaction with the built environment helps us understand the trends in anthropogenic activities and climate crisis. Human thermal comfort is a psychological state of mind that subjectively expresses satisfaction with the surrounding environment. In his seminal work published in 1972, P.O. Fanger introduced statistical models of predicted mean vote (PMV) and percent population dissatisfied (PPD) to predict the thermal comfort of space (16). Human thermal comfort is affected by physiological factors such as occupant metabolic rate, clothing insulation, health conditions, and physical characteristics, including surrounding air temperature, radiant temperature, air velocity, and humidity (17).

Human thermal comfort is highly influenced by the design of

built environments. Building design, construction materials, and mechanical systems create unique indoor environmental conditions, whereas building density, orientation, and amounts of hard-paved surfaces create outdoor conditions in urban areas. These indoor/outdoor microclimates generate distinctive physical conditions that impact the occupants' thermal comfort and health. One such example is the adverse conditions that caused the deaths of 739 residents in the city of Chicago due to a heat wave in 1995 (18). Temperatures in urban areas rose to 106°F due to dense building masses and a lack of ample green spaces. Moreover, in 2002, a heatwave in China led to the death of seven people and 3,500 were hospitalized. This increase in the urban area temperature, compared to the city's synoptic temperature, is called the urban heat island effect. Designing buildings adaptable to changing scenarios is important to maintain comfortable indoor and outdoor conditions (19). We rely heavily on mechanical means to achieve thermal comfort and this is driving greenhouse gas emissions in urban areas. As global temperature rises, buildings use more energy to maintain thermally comfortable indoor environments, leading to increased carbon emissions exacerbating global warming. To break away from this causal nexus of thermal comfort, buildings, and climate crisis, we have to look for novel solutions for achieving thermal comfort while also reducing the built environment's carbon footprint. Researchers are working on innovative ideas to achieve thermal comfort in office buildings. A study reported that mixed-mode office buildings that combine natural ventilation and mechanical cooling reduce energy use by 20-30 percent while maintaining thermal comfort (20). Another study highlighted the importance of personalized control systems in achieving thermal comfort in buildings based on user sensation feedback (21).

Thermal comfort is the body's psychological state in its surrounding environment. The physiological impacts of the built environment can also affect our health and wellness, which are characterized by factors different from thermal comfort. Therefore, it is important to explore how built environments impact health and wellness. This will assist stakeholders in developing strategies that promote positive interaction between humans and the built environment.

Indoor/outdoor environmental quality, health, and wellness

The health and wellness of the urban population are affected by indoor and outdoor environments. As discussed earlier, thermal comfort is affected by the built environment. However, the indoor environment quality (IEQ) also constitutes air quality, lighting, acoustics, and building products that surround us. These factors affect the health and productivity of building occupants and need to be considered when assessing the built environment's impact on health and wellness. Reduced indoor pollutants and optimized comfort parameters can enhance our work performance by 10 percent while reducing absenteeism by 35 percent (22). Meanwhile, improved cognitive function can be achieved by providing optimal IEQ conditions in healthy buildings (23). Outdoor environmental quality (OEQ) is a vital component of urban areas that must be considered when we design our cities. OEQ deals with urban-level aspects such as air pollution, access to green spaces, and biodiversity that are critical to the health and wellness of the urban population.

Rapid urbanization and the burning of fossil fuels have significantly polluted the global air that we depend on for biological functions. OEQ can have a major effect on indoor air quality, particularly on indoor air pollutants. A 60 percent increased risk of developing respiratory illness was associated in children who were exposed to high concentrations of indoor air pollutants (24). Mental health, a critical component of wellness, is also affected by urban environmental conditions. Green spaces provide areas for relaxation and connect the urban population with nature. Reducing cardiovascular disease prevalence and improving mental health outcomes are associated with access to quality green space in urban areas (24).

As extreme weather events coupled with the negative impacts of the climate crisis become more frequent, it is essential to promote a built environment that supports the health and wellness of its occupants. Mental health also needs to be considered when we talk about health and wellness. Having access to quality green spaces is associated with improved mental health and reduced stress in urban populations (25).

All in all, it is recommended to design cities to promote public health and wellness. Although this is the prime focus we strive for, there have been instances when cities are not prepared for unique events such as epidemics or pandemics whose impacts linger longer through mental illnesses. The COVID-19 pandemic is a desolate reminder of the impact of built environment on occupant mental health. Resilience against physiological and mental health requires a better understanding of the psychological health of the people.

Psychology, mental health, and built environment

Physical and mental health are both equally vital parts of overall health. However physical health has always been at the forefront until the recent awareness that mental health is worsening. For instance, the recent pandemic (COVID-19) showed a severe impact on mental health (26). Mental health is the behavioral or cognitive pattern in humans that affects how we act and think, and is triggered by various external factors, which also include the physical environment. Ample research is available on different urban built environments that lead to psychological health conditions such as depression in residential neighborhoods in New York (27), anxiety disorders in the UK residents during COVID (28), loneliness, social isolation, and urban mobility in the Netherlands (29), post-traumatic stress disorder (PTSD) amongst veterans in different indoor spaces (30), schizophrenia and home environment (31), panic disorder recovery and the home environment (32), bipolar disorder and green spaces (33), obsessive-compulsive disorder in children and green spaces (34) and neurodevelopmental disorders and neighborhood conditions (35).

Psychological and physiological relationships in mental health studies

Another focus area in mental health research is the interrelationship between psychology and physiology. Along with psychological wellbeing, the interaction between human behavior and built environments also affects physiological health. For example, one-third of the global population has a sedentary lifestyle, a contributing factor includes the lack of spaces promoting exercise and active behavior. Built environments like office spaces and the influx of television and electronic devices further contributes to the issue. Ultimately, a sedentary lifestyle reduces muscle glucose and protein transporter activities, along with damaging lipid metabolism, to name a few. Furthermore, it decreases blood flow and triggers the sympathetic nervous system, which reduces insulin sensitivity. As a result, people have an increased risk of type 2 diabetes (36). Studies also show the negative impact of sedentary behavior on mental health due to increased stress levels, negative mood, and decreased concentration (37). These same conditions in the longer run lead to psychiatric diseases such as schizophrenia. Thus, psychological wellbeing and physical health are interrelated. Additionally, studies show that a positive psychological outlook could lead to a longer lifespan, slower disease progression, and promote better health (38).

High-rise buildings and mental health

Among various forms of buildings, tall buildings directly and indirectly impact mental health. In this age of increased urbanization and high-rise building developments, research on mental disorders and building design is gaining momentum. One research topic that has been investigated is the relationship between building height in high-rise buildings and its negative influence on women's and children's mental health, especially in low-income families. This is well-established by research studies (6). Findings highlight that a lack of public spaces and a disconnection from the ground level leads to loneliness and emotional strain (39). Another key topic is neighborhood quality and its impact on mental health, especially in poor socioeconomic demographics. It is often found that negative perceptions of neighborhoods due to lack of resources, accessibility, or poor design conditions result in psychological distress.

Figure 3
Urban Plaza in Chicago

Source: Captured by author (Zahida Khan)

Environmental quality and mental health

As much as urban built forms are important, spaces (interior/exterior) and their environmental quality are also critical to mental health as seen in the earlier section. Indoor and outdoor environmental quality can be measured through air quality, lighting levels, thermal comfort conditions, and acoustic levels. Air quality in indoor spaces results from the chemical toxins released through building materials, odors, and furniture. Toxins such as lead, volatile organic compounds, solvents, particulate matter, and pesticides impact cognitive behavior. The study showed that lead exposure in early childhood resulted in psychiatric disorders seen through difficult personality traits such as being antisocial, hyperactivity, consumption of drugs, aggression, and criminal behaviors in adulthood (40). Another study showed an association between depression and common environmental toxicants (41). Unpleasant air pollutants further exacerbate these symptoms and could lead to chronic psychiatric disorders. It is worth noting that many of these pollutants result from climate change events and show an association with health (42). One of the most widely studied pollutants is particulate matter (indoors/outdoors) shows adverse physical and mental health effects. Besides the duration of exposure, PM impacts depend on their size and their ability to cross the blood-brain barrier. This causes inflammation which leads to neurotoxic effects resulting in anxiety, depression, and sometimes neurodevelopmental disorders such as autism or schizophrenia (6)43).

Another metric of environmental quality is sunlight. A lack of sunlight is also associated with disorders such as seasonal affective disorder (SAD) resulting in a sad mood, depression, and low energy (44). The amount of sunlight exposure, mental health, and productivity of office workers are closely linked especially in hospital staff (45). Other studies show a strong relationship between sleep problems and mental health (46). Design measures to mitigate these problems involves making larger windows and/or providing artificial lighting. Similar to lighting levels, thermal comfort levels also impact users' behavioral health (47). One of the direct reactions to unacceptable or adverse thermal conditions is adaptive behaviors by occupants noticed through turning on the thermostat, food/drink intake, changing clothes, turning on fans, etc. It has been recently found that such behaviors have a direct impact on energy consumption and thus indirectly impact climate change. Studies show that a relationship may exist between environmental conditions, behavioral health, and climate change (42). In other words, along with psychological effects, buildings are vital for urban ecology and play a large role in climate change and mental health. A holistic approach is required for built environment design that considers human behavior and environmental factors as metrics for the sustainable development of communities. By doing so, mental disorders could be reduced thus relieving the burden of global disease and premature mortality (48). This also includes addressing the specific needs of varied population to ensure diversity and inclusion.

Diversity and inclusion in mental health and space design

In the context of this research, it can be inferred that occupant behavior (defined as the interaction between an individual's behaviors and built environments) is driven by built environments. Whether it be turning on the thermostat in an office setting when one feels cold or moving to a shaded space when it gets too sunny, people choose (and interact) with environments based on their needs and preferences. However, this phenomenon fails to account for (and accommodate) vulnerable populations —people category defined by the Human Subject Research— which include but are not limited to pregnant women, minors, the elderly, prisoners, students, and persons with diminished mental capacity or cognitive impairments. Mental health conditions are critical to this group. Although a wide range exists in this population, this study attempts to cover selected ones that are directly associated with built environment design and could contribute to highlighting the need for diversity and inclusion in human-eco-centric design that addresses mental health and climate change.

Social group (pregnant women, minors, and the elderly)

The lack of design strategies for social groups namely gender and age has raised the disparity issue in society. Whether the design involves poor IEQ of interior spaces at a building level or a lack of outdoor public spaces at a city level, the population in this group requires special consideration. For instance, the scientific convention for thermal comfort conditions for indoor spaces is based on the average male metabolic rate which overestimates the female thermal demand (49). In office environments, such conditions favor males' productivity and show their female counterparts at a disadvantage leading to lower productivity in women. Since such conventions are adapted to all indoor spaces, the case becomes detrimental to pregnant women as it may aggravate their mental health problems. Studies show that poor air quality results in emotional stress during pregnancy (50). A recent review of the effects of air pollution and children's mental health showed that exposure to chemicals increases children's risk of behavioral problems (51). Acoustic is another environmental factor that is critical to children's mental health. For instance, proximity to airport buildings has shown adverse effects on children's growth and mental health (6). This could be addressed by avoiding residential zones near airport areas. Other city planning policies could include proximity to green parks, which can help reduce mental disorders such as obsessive-compulsive behaviors in children (34). Heat stress due to climate change and poor urban planning has hostile health effects on women, children, and the elderly with a higher risk to the elderly group (52). A study in Korea reported that air pollution may increase depression in the elderly (53). To summarize, built environment design (building/urban planning) has severe health impacts on pregnant women, minors, and the elderly and must be addressed in the design process.

Cognitive and behavioral disorder group

Designing for the population groups showing mental impairment in human cognition and behavioral challenges is often overlooked in conventional design norms. This group includes neurodiverse (autism, attention deficit hyperactivity disorder (ADHD), dyslexia, dyspraxia, and dysgraphia), neurodegenerative (sensory processing disorders such as schizophrenia, dementia, and Parkinson's disease), and personality disorder (rigid behavioral patterns such as antisocial, borderline, histrionic, and narcissistic) individuals (54,55). The US National Research Council reported that more than 28 percent of neurodevelopment disabilities result from environmental conditions (35). For instance, a recent study shows an association between air pollution and neurodevelopmental disorders such as autism in pregnancy exposure (42). Another study show that prolonged exposure to fine and ultrafine particulate matter in children could result in Alzheimers and Parkinsons diseases (56). Besides ensuring proper indoor environmental quality, other design interventions such as flexible furniture layout that allows social interaction and large open sleeping wards in hospitals to reduce loneliness, and provide a homelike environment, good wayfinding, simple floor plans, reduced noise levels, etc. could be very helpful (6).

Since this group interacts and experiences the world differently than neurotypicals, they require spaces that accommodates their needs and allow them to thrive. Flexibility and choices are two powerful design strategies that could be beneficial in this scenario. It is important to design accommodations that allow inclusion and break structural disparities.

Prison populations

Statistical reports from the US Department of Justice show that prison populations are extremely vulnerable to mental disorders (57). Prison designs are critical since the prime aim of this facility is to reinstate them into society (58). It can either aggravate or alleviate their mental conditions. Studies show that the prison environment is often neglected and is more focused on confinement than on rehabilitation. This has increased mental disorders in prison populations (59). However, there are strategies to address this issue. For instance, proximity to nature through a window view helps to counter-act the sanitary environments in correctional facilities (60). Mental stress, anxiety, and negative mood can be addressed through stress reduction by involving exposure to the natural landscape or green spaces (61). Another important factor in prison design is to ensure personal space allocation to each prisoner and to not overcrowd the facility in order to reduce anxiety and stress.

It is recommended that built environments should consider strategies around passive design to address mental health. Advocates in this area outline design strategies such as improved daylighting and artificial lighting, good air quality, greenery, proper acoustics, appropriate density for personal space, and appealing sensory designs that contribute significantly to health design.

Design strategies to combat climate change

Investing in healthy indoor and outdoor environments is necessary to achieve a sustainable and resilient urban future. The built environment not only affects the health and wellness of its occupants but also impacts the stability of the planet. Although the complex interaction among anthropogenic activities, planetary boundaries, and human health/wellness must be explored to understand the detrimental impacts of the climate crisis, a few strategies that have been attempted to address these interactions are discussed below.

Circular approach

The built environment influences the global climate crisis through anthropogenic activities, accounting for 37 percent of global carbon emissions and the use of resources. Carbon emissions and their adverse effects on the planet can be mitigated through design strategies such as energy efficiency, sustainable materials choices, resilient urban planning, and integrating natural systems. Moreover, with a unique approach referred to as the "circular design thinking" approach, it is possible to create products and systems that, at the end of their life cycles can either biodegrade harmlessly or can be fully recycled into new high-quality products generating minimal waste and negative environmental impact (62).

Sustainable urbanism approach

Sustainable urbanism is another design approach that tackles the impacts of the built environment at a larger scale. Walkable communities that reduce reliance on cars can significantly lower emissions associated with transportation (63). Sustainable urbanism promotes higher densities and mixed-use development to reduce energy consumption through public transit, biking, and walking (64). Ian McHarg sheds light on the concept of ecological planning in urban development in one of its seminal works. McHarg proposes integrating natural systems, such as wetlands and green spaces within built environments, that lead to resilient cities regarding flooding and other climate-related events, particularly heat waves (65). The contemporary sponge cities of China are a great example of integrating a variety of green urban spaces to prevent flooding.

Material choice

Material choices and energy efficiency play a central role in sustainable design of the built environment. Material choices can dramatically affect the sustainability across the life cycle of a building. Environmental impacts characterize construction materials and must be carefully considered when designing built environments. Using local, renewable, low-energy materials, and vernacular construction techniques developed specifically for local microclimates can significantly reduce the carbon footprint (66).

Passive/active building systems

Passive and active design strategies can also reduce energy consumption in the built environment, making it more energy efficient. Dependence on artificial heating and cooling can be reduced through optimized building geometry to allow for natural light, solar heat gain, and natural ventilation. Active strategies involve incorporating renewable energy systems such as solar panels and wind turbines. Innovation in manufacturing and resourcing has made using renewable energy systems increasingly possible and cost-effective for new projects and retrofits. The Bullitt Center in Seattle, known as the greenest commercial building in the world, shows one way these strategies can be put into practice. Off-grid building incorporates state-of-the-art technologies and materials to strive toward net-positive energy, water, and waste systems (67).

Biophilic design

Another popular strategy incorporating natural elements in the built environment is known as biophilic design which has been proven to impact human wellbeing and environmental sustainability positively. Green roofs, living walls, and urban forests have the potential to enhance urban biodiversity, improve air quality, and moderate the heat island effect in urban areas (68). The Bosco Verticale of Milan is an example of an urban vertical forest hosting over 20,000 plants.

Tools/frameworks

Integrating green building certification and rating systems like LEED and Green Globes in professional practice in the United States ensures the fostering of sustainable design practices (69). Such systems provide designers and stakeholders with tools that offer assessment frameworks for the sustainability impacts of buildings while functioning to drive innovation by setting high-performance goals. Such certifications facilitate energy-efficient systems, sustainable materials, and waste reduction practices. On a larger scale, policy and global initiatives drive the practice toward sustainability in the built environment. The United Nations Environment Programme 2009 report provides a valuable overview of the building sector's contribution to global emissions while outlining policy measures to ensure that buildings become more energy-efficient and sustainable. Targets such as net-zero emissions can be achieved by taking significant steps to design and construct sustainable buildings (70).

Assessing the environmental impact of buildings and urban environments is a complex process that requires a sophisticated approach. Several frameworks and assessment tools have been developed over the years that assist architects, designers, and stakeholders in quantifying the performance of the built environment. These tools encompass a variety of building types as well as distinct building phases (71). With the advent of the 21st-century digital age, advanced computational means, such as parametric and algorithmic design, help architects and planners predict, quantify, and optimize building performance for energy efficiency and environmental impact. These tools have assisted stakeholders in complex decision-making by enabling accurate simulations of the built environment and assessing the future ecological implications (72).

Integrating these strategies in the design process often requires a holistic approach considering technological innovation, ecological sensitivity, and socioeconomic factors. Sustainable thinking must be at the forefront when anthropogenic activities and their environmental impacts are considered. However, simply maintaining the current conditions of the earlier discussed planetary boundaries is insufficient to mitigate the adverse effects of the built environment. The built environment must begin to impact its users and the planet positively. While sustainability is often integrated into creating a place, it can only be considered genuinely regenerative if it positively impacts its surroundings (73). Regenerative design looks beyond sustainability by comprehensively exploring the relationship between human and natural systems. It is a "whole system thinking" that strives to create a resilient and equitable design that promotes the needs of society while maintaining the integrity of nature (74).

Conclusion

This study explores the relationship between building stocks, climate, and mental health while understanding human interaction with built environments. The findings highlight the current trend in this topic and recommend an in-depth interdisciplinary study aiming for further innovation in design strategies and approaches catering to human eco-centric design. Current and future generations of designers, builders, policy-makers, and stakeholders must work together to promote sustainable strategies towards resilient and regenerative urban ecology.

Additionally, it is worth highlighting that mental health is an important topic in the "built environment" policy-level areas, such as the SDGs. Although it is getting wider attention through initiatives such as the yearly UN campaigns on UN Mental Health Day (75) and the WHO's special initiative for mental health (76), there is limited discussion on sustainable built environment policies. Studies show that mental health has been of prime concern recently because of the pandemic (COVID-19) (77). It is recommended that SDG-3 should also consider mental health in its vision of "good health and well-being". It might not be pressing today, but it will be by 2050 as we face the challenges of epidemics and pandemics along with urbanization and the climate crisis.

References

1.United Nations. The world's cities in 2018. UN Statistical Papers - United Nations (Ser. A), population and vital statistics report. United Nations. https://www.un-ilibrary.org/content/books/9789210582766

2.UN DESA. The sustainable development goals report 2023: Special Edition. United Nations. 2023. https://www.un-ilibrary.org/content/books/9789210024914

3.Becerik-Gerber B, Lucas G, Aryal A, Awada M, Bergés M, Billington S, et al. The field of human building interaction for convergent research and innovation for intelligent built environments. Scientific Reports. 2022 Dec 21;12(1):22092. https://www.nature.com/articles/s41598-022-25047-y

4.ASHRAE 62. Ventilation and acceptable indoor air quality. Atlanta, GA. ASHRAE. 2023. https://store.accuristech.com/ash rae/standards/ashrae-62-1-2022?product_id=2501063

5.Aljunaidy MM & Adi MN. Architecture and mental disorders: A systematic study of peer-reviewed literature. HERD. 2021 Jul 1;14(3):320–30. https://pubmed.ncbi.nlm.nih.gov/33356588/

6.Evans GW. The built environment and mental health. Journal of Urban Health. 2003 Dec; 80:536–55. https://pubmed.ncbi.n lm.nih.gov/14709704/

7.WHO. Mental disorders. World Health Organization. 2022. https://www.who.int/news-room/fact-sheets/detail/mental-disorders

8.Steinfeld E & Maisel J. Universal design: Creating inclusive environments. 1st edition. New Jersey: John Wiley & Sons, Ltd. 2012. https://www.wiley.com/en-us/Universal+Design%3A+Creating+Inclusive+Environments-p-9780470399132

9.Cabeza L, Bai Q, Bertoldi P, Kihila JM, Lucena A, Mata E, et al. Chapter 9: Buildings. In: IPCC sixth assessment report. Cambridge, UK and New York USA: Cambridge University Press. 2022. https://www.ipcc.ch/report/ar6/wg3/chapter/chapter-9/

10.Dawson A. Extinction: A radical history. OR Books. 2016. https://www.jstor.org/stable/j.ctv62hf5h

11.Steffen W, Richardson K, Rockström J, Cornell SE, Fetzer

I, Bennett EM, et al. Planetary boundaries: Guiding human development on a changing planet. Science. 2015 Feb 13;347(62 23):1259855. https://www.science.org/doi/10.1126/science.12 59855

12.Wallace-Wells D. The uninhabitable Earth: Life after warming. Tim Duggan Books. 2019. https://www.amazon.in/Uninh abitable-Earth-Life-After-Warming/dp/0525576703

13.UN EP. 2023 Global status report for buildings and construction: Beyond foundations - Mainstreaming sustainable solutions to cut emissions from the buildings sector. United Nations Environment Programme. 2024. https://wedocs.unep. org/20.500.11822/45095

14.WMO. State of the global climate 2023. Geneva: WMO. 2024 https://library.wmo.int/records/item/68835-state-of-the-g lobal-climate-2023

15.WEF. The Global risks report 2024. Switzerland: World Economic Forum. 2024 Jan. https://www3.weforum.org/docs/ WEF_The_Global_Risks_Report_2024.pdf

16.Shaw EW. Thermal comfort: analysis and applications in environmental engineering, by P. O. Fanger. 244 pp. DANISH TECHNICAL PRESS. Copenhagen, Denmark. 1970. Danish Kr. 76, 50. Royal Society of Health Journal. 1972 Jun 1;92(3):164–164. https://www.semanticscholar.org/paper/T hermal-Comfort%3A-analysis-and-applications-in-by-P.-Shaw/dd5d96a1d6cd1001e4fd62ef6b94a9466905dd36

17.Parsons K. Human thermal environments: The effects of hot, moderate, and cold environments on human health, comfort, and performance. Third Edition. 3rd ed. Boca Raton: CRC Press. 2014. https://www.taylorfrancis.com/books/mono/10.1201/b1 6750/human-thermal-environments-ken-parsons

18.Dematte JE, O'Mara K, Buescher J, Whitney CG, Forsythe S, McNamee T, et al. Near-fatal heat stroke during the 1995 heat wave in Chicago. Annals of Internal Medicine. 1998 Aug;129(3):173–81. https://pubmed.ncbi.nlm.nih.gov/969672 4/

19.Nicol F, Humphreys M &Roaf S. Adaptive thermal comfort: Principles and practice. London: Routledge. 2012; 208. https://www.routledge.com/Adaptive-Thermal-Comfort-Pri nciples-and-Practice/Nicol-Humphreys-Roaf/p/book/9780 415691598?srsltid=AfmBOoo7w7ICiCH344tir3BSo3ziDRL0qa Q0N9HqYwuCt8BsSUg8M5Hs

20.Kim J, Tartarini F, Parkinson T, Cooper P & de Dear R. Thermal comfort in a mixed-mode building: Are occupants more adaptive? Energy and Buildings. 2019 Nov 15; 203:109436. https://ro.uow.edu.au/articles/journal_contribution/Therma l_comfort_in_a_mixed-mode_building_Are_occupants_ more_adaptive_/27764193

21.Nouvel R & Alessi F. A novel personalized thermal comfort control, responding to user sensation feedbacks. Building Simulation. 2012 Sep 1;5(3):191–202. https://link.springer.co m/article/10.1007/s12273-012-0076-5

22.Wargocki P & Wyon DP. The effects of moderately raised classroom temperatures and classroom ventilation rate on the performance of schoolwork by children (RP-1257). HVAC&R Research. 2007 Mar 1;13(2):193–220. https://orbit.dtu.dk/en /publications/the-effects-of-moderately-raised-classroom-temperatures-and-class

23.Grimm NB. Urban ecology: what is it and why do we need it? Urban ecology: its nature and challenges. CABI Digital Library. 2020 Jan;1–14. https://www.cabidigitallibrary.org/doi/10.107 9/9781789242607.0001

24.Lanphear BP, Aligne CA, Auinger P, Weitzman M & Byrd RS. Residential exposures associated with asthma in US children. Pediatrics. 2001 Mar 1;107(3):505–11. https://pubmed.ncbi.nl m.nih.gov/11230590/

25.van den Berg M, Wendel-Vos W, van Poppel M, Kemper H, van Mechelen W & Maas J. Health benefits of green spaces in the living environment: A systematic review of epidemio-logical studies. Urban Forestry & Urban Greening. 2015 Jan 1;14(4):806–16. https://www.sciencedirect.com/science/artic le/pii/S1618866715001016

26.WHO. Mental Health and COVID-19: Early evidence of the pandemic's impact. WHO. 2022 Mar. https://www.who.int/ publications/i/item/WHO-2019-nCoV-Sci_Brief-Mental_he alth-2022.1

27.Galea S, Ahern J, Rudenstine S, Wallace Z & Vlahov D. Urban built environment and depression: A multilevel analysis.

Journal of Epidemiology and Community Health. 2005 Oct; 59(10):822–7. https://pubmed.ncbi.nlm.nih.gov/16166352/

28.Xiao J, Zhao J, Luo Z, Liu F & Greenwood D. The impact of built environment on mental health: A COVID-19 lockdown perspective. Health & Place. 2022 Sep 1; 77:102889. https://www.sciencedirect.com/science/article/pii/S1353829222001502

29.van den Berg P, Kemperman A, de Kleijn B, Borgers A. Ageing and loneliness: The role of mobility and the built environment. Travel Behaviour and Society. 2016 Sep 1; 5:48–55. https://www.sciencedirect.com/science/article/abs/pii/S2214367X15000101

30.Nuamah J, Rodriguez-Paras C & Sasangohar F. Veteran-centered investigation of architectural and space design considerations for post-traumatic stress disorder (PTSD). HERD. 2021 Jan 1;14(1):164–73. https://pubmed.ncbi.nlm.nih.gov/32441151/

31.Karol E & Smith D. Design considerations for residents with impeded cognitive functioning: conversations with people with schizophrenia. Sustainability. 2021 Jan;13(14):7733. https://www.mdpi.com/2071-1050/13/14/7733

32.Espeso CSR. From safe places to therapeutic landscapes: The role of the home in panic disorder recovery. Wellbeing, Space and Society. 2022 Jan 1; 3:100108. https://www.research.ed.ac.uk/en/publications/from-safe-places-to-therapeutic-landscapes-the-role-of-the-home-i

33.Chang HT, Wu CD, Wang JD, Chen PS & Su HJ. Residential green space structures are associated with a lower risk of bipolar disorder: A nationwide population-based study in Taiwan. Environmental Pollution. 2021 Aug 15; 283:115864. https://researchoutput.ncku.edu.tw/en/publications/residen tial-green-space-structures-are-associated-with-a-lower-ri

34.Ezpeleta L, Navarro JB, Alonso L, de la Osa N, Ambrós A, Ubalde M, et al. Greenspace exposure and obsessive-compulsive behaviors in schoolchildren. Environment and Behavior. 2022 Jun 1;54(5):893–916. https://portalrecercat.ua b.cat/en/publications/greenspace-exposure-and-obsessive-compulsive-behaviors-in-schoolc

35.Santiago AM, Berg KA & Leroux J. Assessing the impact of neighborhood conditions on neurodevelopmental disorders during childhood. International Journal of Environment Research and Public Health. 2021 Aug 27;18(17):9041. https://pm c.ncbi.nlm.nih.gov/articles/PMC8430861/

36.Park JH, Moon JH, Kim HJ, Kong MH & Oh YH. Sedentary lifestyle: Overview of updated evidence of potential health risks. Korean J Fam Med. 2020 Nov;41(6):365–73. https://pmc.ncbi. nlm.nih.gov/articles/PMC7700832/

37.Falk GE, Mailey EL, Okut H, Rosenkranz SK, Rosenkranz RR, Montney JL, et al. Effects of sedentary behavior interventions on mental well-being and work performance while working from home during the COVID-19 pandemic: A pilot randomized controlled trial. International Journal of Environment Research and Public Health. 2022 May 24;19(11):6401. https://pubmed.ncbi.nlm.nih.gov/35681986/

38.Hernandez R, Bassett SM, Boughton SW, Schuette SA, Shiu EW & Moskowitz JT. Psychological well-being and physical health: Associations, mechanisms, and future directions. Emotion Review. 2018 Jan 1;10(1):18–29. https://pubmed.ncbi.nlm.nih.gov/36650890/

39.Larcombe DL, van Etten E, Logan A, Prescott SL & Horwitz P. High-rise apartments and urban mental health—historical and contemporary views. Challenges. 2019 Dec;10(2):34. https://www.mdpi.com/2078-1547/10/2/34

40.Reuben A, Schaefer JD, Moffitt TE, Broadbent J, Harrington H, Houts RM, et al. Association of childhood lead exposure with adult personality traits and lifelong mental health. JAMA Psychiatry. 2019 Apr;76(4):418–25. https://jamanetwork.com/journals/jamapsychiatry/fullarticle/2720691

41.Guo J, Garshick E, Si F, Tang Z, Lian X, Wang Y, et al. Environmental toxicant exposure and depressive symptoms. JAMA Network Open. 2024 Jul 3;7(7):e2420259. https://jamanetwork.com/journals/jamanetworkopen/fullarticle/2820702

42.Cuijpers P, Miguel C, Ciharova M, Kumar M, Brander L, Kumar P, et al. Impact of climate events, pollution, and green spaces on mental health: an umbrella review of meta-analyses. Psychological Medicine. 2023 Feb;53(3):638–53. https://pubmed.ncbi.nlm.nih.gov/36606450/

43.Bhui K, Newbury JB, Latham RM, Ucci M, Nasir ZA, Turner B, et al. Air quality and mental health: Evidence, challenges and future directions. BJPsych Open. 2023 Jul 5;9(4):e120. https://p ubmed.ncbi.nlm.nih.gov/37403494/

44.Melrose S. Seasonal affective disorder: An overview of assessment and treatment approaches. Depression Research and Treatment. 2015;2015:178564. https://pubmed.ncbi.nlm. nih.gov/26688752/

45.Wang J, Wei Z, Yao N, Li C & Sun L. Association between sunlight exposure and mental health: Evidence from a special population without sunlight in work. Risk Management and Healthcare Policy. 2023 Jun 14;16:1049–57. https://pmc.ncbi. nlm.nih.gov/articles/PMC10277019/

46.Raza A, Partonen T, Hanson LM, Asp M, Engström E, Wester-erlund H, et al. Daylight during winters and symptoms of depression and sleep problems: A within-individual analysis. Environment International. 2024 Jan 1;183:108413. https://pu bmed.ncbi.nlm.nih.gov/38171042/

47.Yang Y, Bai Y, Zhang R, Zhu X. The effect of thermal envi-ronment on stress and thermal comfort of college students under acute stress. Indoor and Built Environment. 2022 Nov 1;31(9):2226–39. https://journals.sagepub.com/doi/abs/10.11 77/1420326X221086193

48.Charlson F, Ali S, Benmarhnia T, Pearl M, Massazza A, Augustinavicius J, et al. Climate change and mental health: A scoping review. International Journal of Environmental

Research and Public Health. 2021 Apr 23;18(9):4486. https://p
mc.ncbi.nlm.nih.gov/articles/PMC8122895/

49.Kingma B, van Marken Lichtenbelt W. Energy consumption
in buildings and female thermal demand. Nature Climate
Change. 2015 Dec;5(12):1054–6. https://econpapers.repec.o
rg/RePEc:nat:natcli:v:5:y:2015:i:12:d:10.1038_nclimate2741

50.Lin Y, Zhou L, Xu J, Luo Z, Kan H, Zhang J, et al. The impacts
of air pollution on maternal stress during pregnancy. Sci Rep.
2017 Jan 18;7(1):40956. https://www.nature.com/articles/sre
p40956

51.James AA & O'Shaughnessy KL. Environmental chemical
exposures and mental health outcomes in children: a narrative review of recent literature. Front Toxicol. 2023 Nov
30;5:1290119. https://pubmed.ncbi.nlm.nih.gov/38098750/

52.Kovats RS, Hajat S. Heat stress and public health: A
critical review. Annual Review of Public Health. 2008 Apr
1;29(29):41–55. https://pubmed.ncbi.nlm.nih.gov/18031221/

53.Lim YH, Kim H, Kim JH, Bae S, Park HY & Hong YC. Air pollution and symptoms of depression in elderly adults. Environmental Health Perspectives. 2012 Jul;120(7):1023–8. https://p
ubmed.ncbi.nlm.nih.gov/22514209/

54.BSI. PAS 6463:2022 Design for the mind. Neurodiversity
and the built environment. Guide. In London: British Standard
Institute. 2022. https://knowledge.bsigroup.com/products/
design-for-the-mind-neurodiversity-and-the-built-enviro

nment-guide?version=standard

55.Getten J. Neurodiversity vs. personality disorders: Navigating the nuances for better understanding. Behavioral Health Consulting Solutions. 2024. https://www.bhcsmt.com/blog/neurodiversity-vs-personality-disorders

56.Calderón-Garcidueñas L, Calderón-Garcidueñas A, Torres-Jardón R, Avila-Ramírez J, Kulesza RJ & Angiulli AD. Air pollution and your brain: what do you need to know right now. Primary Health Care Research & Development. 2015 Jul;16(4):329-45.https://www.cambridge.org/core/journals/primary-health-care-research-and-development/article/air-pollution-and-your-brain-what-do-you-need-to-know-right-now/D60A0C1A1801217E354311DD025992C9

57.James D & Glaze L. Mental health problems of prison and jail inmates. US Department of Justice, Office of Justice Programs; 2006 Dec. https://bjs.ojp.gov/content/pub/pdf/mhppji.pdf

58.Cunha O, Castro Rodrigues A de, Caridade S, Dias AR, Almeida TC, Cruz AR, et al. The impact of imprisonment on individuals' mental health and society reintegration: study protocol. BMC Psychol. 2023 Jul 25;11:215. https://bmcpsychology.biomedcentral.com/articles/10.1186/s40359-023-01252-w

59.Gómez-Figueroa H & Camino-Proaño A. Mental and behavioral disorders in the prison context. Revista Espanola de Sanidad Penitenciaria. 2022 Oct 6;24(2):66-74. https://pmc.ncbi.nlm.nih.gov/articles/PMC9578298/

60.Jewkes Y. Just design: Healthy prisons and the architecture of hope. Australian & New Zealand Journal of Criminology. 2018 Sep 1;51(3):319–38. https://researchportal.bath.ac.uk/en/publ ications/just-design-healthy-prisons-and-the-architecture -of-hope

61.Moran D, Jones PI, Jordaan JA & Porter AE. Nature contact in the carceral workplace: Greenspace and staff sickness absence in prisons in England and Wales. Environment and Behavior. 2022 Feb 1;54(2):276–99. https://journals.sagepub.com/doi/ 10.1177/00139165211014618

62.Braungart M & McDonough W. Cradle to cradle. Penguin. 2009. https://www.penguin.co.uk/books/405817/cradle-to-cradle-by-michael-braungart-and-william-mcdonough/97 80099535478

63.Farr D. Sustainable urbanism: Urban design with nature. Wiley. 2007. https://www.wiley.com/en-us/Sustainable+Urb anism%3A+Urban+Design+With+Nature-p-9780471777519

64.Ewing R & Rong F. The impact of urban form on U.S. residential energy use. Housing Policy Debate. 2008 Jan 1;19(1):1–30. https://www.researchgate.net/publication/238587126_The_ Impact_of_Urban_Form_on_US_Residential_Energy_Us e

65.McHarg I. Design with nature, 25th anniversary edition. Wiley. 1995. https://www.wiley.com/en-us/Design+with+N ature%2C+25th+Anniversary+Edition-p-9780471114604

66.Berge B. The ecology of building materials. 2nd ed. Routledge; 2009. https://www.routledge.com/The-Ecology-of-B uilding-Materials/Berge/p/book/9781856175371

67.Thomas MA. The Greenest Building: How the bullitt center changes the urban landsc. International Living Future Institute. 2015. https://store.living-future.org/products/the-greenest-building-how-the-bullitt-center-changes-the-urban-land scape

68.Sanesi G, Colangelo G, Lafortezza R, Calvo E & Davies C. Urban green infrastructure and urban forests: A case study of the Metropolitan Area of Milan. Landscape Research. 2017 Feb 17;42(2):164–75. https://ideas.repec.org/a/taf/clarxx/v42y20 17i2p164-175.html

69.Smith T, Fischlein M, Suh S & Huelman P. Green building rating systems: A comparison of the LEED and green globes systems in the US. prepared for the Western Council of Industrial Workers, September. 2006 Jan 1. https://www.researchga te.net/publication/230838441_Green_Building_Rating_Sys tems_a_Comparison_of_the_LEED_and_Green_Globes_ Systems_in_the_US

70.UN EP. 2019 Global status report for buildings and construction sector. UNEP - UN Environment Programme. UN Environment and International Energy Agency. 2019 Jan. https://www.unep.org/resources/publication/2019-global-s tatus-report-buildings-and-construction-sector

71.Haapio A & Viitaniemi P. A critical review of building environmental assessment tools. Environmental Impact Assessment Review. 2008 Oct 1;28(7):469–82. https://citeseerx.ist.psu.e du/document?repid=rep1&type=pdf&doi=80fd617c4b4a2d6b bdf57712bf7786c1873ea973

72.Caetano I, Santos L, Leitão A. Computational design in architecture: Defining parametric, generative, and algorith-mic design. Frontiers of Architectural Research. 2020 Jun 1;9(2):287–300. https://www.sciencedirect.com/science/art icle/pii/S2095263520300029

73.Mang P & Reed B. Regenerative development and design. In: Loftness V, editor. Sustainable built environments. New York, NY: Springer US. 2020 https://doi.org/10.1007/978-1-0716-0 684-1_303

74.Lyle JT. Regenerative design for sustainable development. Wiley. 1996. https://www.wiley.com/en-us/Regenerative+De sign+for+Sustainable+Development-p-9780471178439

75.United Nations. World mental Health Day. United Nations. https://www.un.org/en/healthy-workforce/world-mental-h ealth-day

76.Kestel D. The state of mental health globally in the wake of the COVID-19 pandemic and progress on the WHO Special Initiative for Mental Health (2019-2023). United Nations. 2022. https://www.un.org/en/un-chronicle/state-mental-health-globally-wake-covid-19-pandemic-and-progress-who-spe cial-initiative

77.Stock S, Bu F, Fancourt D & Mak HW. Longitudinal associations between going outdoors and mental health and wellbeing during a COVID-19 lockdown in the UK. Sci Rep. 2022 Jun 22;12(1):10580. https://pubmed.ncbi.nlm.nih.gov/35732816/

7

Health and Bone Health in the Context of Climate Change and Global Warming

Abstract

This chapter explores the multifaceted impact of climate change on bone health. It highlights the often-overlooked connection between global warming and the deterioration of bone health due to changing environmental conditions. Rising temperatures and extreme weather events affect nutritional availability, reducing the intake of essential nutrients like calcium and vitamin D, which are critical for bone health. Additionally, physical activity, a key factor in maintaining bone density, is compromised by increasing heat, particularly in vulnerable populations such as older people. The chapter also delves into the socioeconomic inequalities exacerbated by climate change noting that poorer communities are disproportionately affected, leading to a higher prevalence of bone-related diseases like osteoporosis. Mitigation strategies, including urban planning innovations, technological advancements, and targeted public health campaigns are discussed as necessary measures to combat these challenges. The chapter concludes with a call for coordinated global efforts to address the intersection of bone health and climate change advocating for policy development, education, and innovation to safeguard bone health in a warming world.

Keywords: Bone health, global warming, socioeconomic inequalities, aging, climate change

Author

Dr Taher Mahmud, Consultant Rheumatologist, Co-Founder of London Osteoporosis Clinic; Trustee of Global Osteoporosis Foundation; Director YouOptimised [tm@drmahmud.com]

Introduction

As the world confronts the escalating challenges posed by climate change, its effects on human health have become increasingly evident. Bone health is a critical issue often overlooked among the myriad health concerns. This chapter explores how global warming, characterised by rising temperatures, changing weather patterns, and the increased frequency of extreme weather events, complicates efforts to maintain bone health—a fundamental aspect of human wellbeing (1).

The physiology of bone health

Bones are dynamic, living tissues that provide structure to the body, protect internal organs, and enable movement. They also serve as a reservoir for essential minerals like calcium and phosphorus. Various factors, including genetics, nutrition, physical activity, and hormonal changes, influence the health of our bones. Maintaining strong bones throughout life is essential for preventing osteoporosis, a disease characterised by weakened bones and an increased risk of fractures.

The impact of climate change on nutrition and bone health

One of the most direct ways climate change affects bone health is through its impact on nutrition. Global warming is altering the availability and quality of food, with significant consequences for the intake of bone-essential nutrients such as calcium and vitamin D. As temperatures rise, agricultural productivity is declining in many regions, leading to reduced yields of crops that are vital for bone health. For example, extreme heat and droughts are compromising the availability of leafy greens and dairy products, which are key sources of calcium (2).

In addition to reduced food production, extreme weather events—such as floods and hurricanes—disrupt food supply chains, leading to periods of scarcity that exacerbate nutritional deficiencies. These challenges are particularly severe in regions already facing poverty and limited access to healthcare, thereby worsening inequalities in bone health outcomes. In communities where access to diverse, nutrient rich foods is already limited, the effects of climate change can lead to increased prevalence of conditions like osteoporosis and other bone-related diseases (3)(4).

Physical activity and bone health in a warmer world

Physical activity is a cornerstone of maintaining bone health, as weight-bearing and resistance exercises help build and maintain bone density (5). However, as global temperatures rise, the feasibility of engaging in outdoor physical activities is becoming increasingly compromised, especially in regions experiencing extreme heat. For instance, cities like Dubai and Mumbai are already grappling with temperatures exceeding 50°C, making outdoor activities uncomfortable and potentially hazardous (6).

Reduced physical activity due to extreme heat has severe implications for bone health. In older adults at risk of osteoporosis, decreased mobility can significantly weaken bones. The situation is compounded by many people needing access to air-conditioned indoor spaces to exercise safely. This chapter discusses potential mitigation strategies, such as developing indoor exercise programs and creating community-based initiatives to keep people active despite the challenging climate (7).

Socioeconomic inequalities and bone health

The intersection of climate change, health, and socioeconomic inequality is a critical theme in understanding the broader impacts on bone health. Wealth disparities significantly influence the ability of individuals and communities to adapt to the health challenges posed by climate change. In many Gulf countries, for example, extreme heat exacerbates existing inequalities, with wealthier individuals able to mitigate the effects of extreme temperatures through access to air-conditioned environments and indoor recreational facilities. In contrast, poorer populations, particularly migrant workers, are left to endure harsh conditions with limited access to adequate nutrition, healthcare, and safe spaces for physical activity.

These disparities are not limited to the Gulf region. Globally, similar patterns are emerging, where poorer communities face more significant risks to their bone health due to climate change. In regions where access to healthcare and nutritious food is already limited, the additional stress of a warming climate further entrenches health inequalities, leaving the most vulnerable populations at an increased risk of bone-related conditions.

Climate change, aging, and bone health

The ageing population is particularly vulnerable to climate change's impacts on bone health. As people age, their bones naturally lose density, making them more susceptible to fractures. Global warming's effects exacerbate this risk. Older adults are more likely to experience heat-related illnesses which can further compromise their bone health.

In regions experiencing extreme heat, older people are often confined indoors, leading to decreased physical activity, essential for maintaining bone strength (8). Additionally, the physiological effects of heat stress can exacerbate existing health conditions leading to a vicious cycle of declining health. This section explores strategies to protect the ageing population, including tailored nutritional guidance, safe physical activity programs, and community support systems designed to help older adults maintain their bone health in a changing climate.

Nutrition interventions for bone health

Addressing the nutritional challenges climate change poses is crucial for maintaining bone health. Innovative agricultural practices and biofortification techniques are being explored to ensure consistent access to bone-healthy foods in a warming world. For example, climate-resilient crops that can withstand extreme weather conditions are being developed to secure the supply of essential nutrients like calcium and vitamin D.

In addition to improving food security, dietary supplementation is emerging as a vital strategy for mitigating the impact of climate-induced nutritional deficiencies in bone health. Public health campaigns are also crucial in raising awareness about the importance of nutrition for bone health, particularly in communities most vulnerable to climate change (9).

Technological and structural adaptations

Urban planning and architectural innovations are increasingly recognized as essential tools in supporting bone health in the context of climate change. Cities can be designed to encourage physical activity even in extreme climates, with features such as air-conditioned walkways, community exercise spaces, and green building technologies that reduce indoor temperatures (10).

Technological advancements, such as wearable devices that monitor bone health and encourage safe physical activity, also contribute to mitigating climate-related health risks. This section discusses the potential of these technologies to empower individuals to take control of their bone health, even as the external environment becomes increasingly challenging.

Policy and advocacy for bone health in the climate era

Effective policy and advocacy are essential for addressing the complex challenges of climate change and bone health. Governments and international organizations must prioritize developing and implementing policies that protect the most vulnerable populations from the health impacts of global warming. This includes ensuring access to nutritious foods, safe spaces for physical activity, and adequate healthcare services.

Advocacy efforts are particularly important in raising awareness about the risks posed by climate change to bone health and for pushing for the inclusion of bone health considerations in broader public health and climate policies. Healthcare providers also play a critical role in this effort by educating patients and communities about the importance of maintaining bone health in a changing climate.

Conclusion

As the planet continues to warm, the challenges to maintaining bone health will become increasingly severe. However, with proactive measures and a commitment to addressing these challenges head-on, it is possible to protect bone health even with climate change. This chapter has explored the complex interplay between global warming, nutrition, physical activity, socioeconomic inequality, and bone health, highlighting the urgent need for coordinated global efforts to address these issues.

Empowerment through knowledge, innovation in nutrition and urban planning and robust advocacy for vulnerable populations ensure that everyone can maintain strong bones and overall health in a rapidly changing world regardless of circumstances. By recognizing the critical importance of bone health in the context of climate change, we can take meaningful steps to safeguard our health and wellbeing in the years to come.

References

1. Intergovernmental Panel on Climate Change. Climate change 2021: The physical science basis. Cambridge University Press. 2021. https://www.ipcc.ch/report/ar6/wg1/
2. World Health Organization. Climate Change and Health. WHO. 2021. https://www.who.int/news-room/fact-sheets/detail/climate-change-and-health
3. Cosman F, Beur SJ, LeBoff MS, Lewiecki EM, et. al. Clinician's guide to prevention and treatment of osteoporosis. Osteoporosis Int. 2014 Aug 15; 25(10):2359-2381. https://www.ncbi.nlm.nih.gov/pmc/articles/PMC417657

3/

4. Haines, A., & Ebi, K. L. The imperative for climate action to protect health. New England Journal of Medicine. 2019; 380(3): 263-273. https://www.nejm.org/doi/full/10.1056/NEJMra1807873

5. Crimmins, A., Balbus, J., Gamble, J. L., Beard, C. B., Bell, J. E., Dodgen, D., et. al. The impacts of climate change on human health in the United States: A scientific assessment. U.S. Global Change Research Program. 2016. https://health2016.globalchange.gov/

6. Financial Times. Learning to live with 50C temperatures. Financial Times. 2024. FT.com.

7. Public Health England. Health impacts of climate change in the UK: An update of the evidence base. UK Health Security Agency. 2018. https://assets.publishing.service.gov.uk/media/659ff6a93308d200131fbe78/HECC-report-2023-overview.pdf

8. London Osteoporosis Clinic. Climate change and bone health: A new challenge. London Osteoporosis Clinic. 2022. LondonOsteoporosisClinic.com.

9. Hernlund, E., Svedbom, A., Ivergård, M., Compston, J., Cooper, C., Stenmark, J., et. al.. Osteoporosis in the European Union: Medical management, epidemiology and economic burden. Archives of Osteoporosis. 2013; 8(1-2), 136. https://pubmed.ncbi.nlm.nih.gov/24113837/

10. Rocque, R.J., Beaudoin, C., Ndjaboue R., Cameron, L, et. al. Health effects of climate change: An overview of systematic reviews. British Medical Journal. 2021. 10.1136/bmjopen-2020-046333

8

Acronyms

ADHD Attention Deficit Hyperactivity Disorder
ASF Superior Audit Office of the Federation
BMGF Bill & Melinda Gates Foundation
BOD Biological Oxygen Demand
BPA Bisphenol A
CBD Convention on Biological Diversity
CDC Center for Disease Control and Prevention
CFD Computer Fluid Dynamics
CO_2 Carbon dioxide
COD Chemical Oxygen Demand
COFEPRIS Federal Commission for Protection Against Health Risks
CONAGUA National Water Commission
ECs Emerging Contaminants
EDCs Endocrine Disrupting Chemicals
FAO Food and Agriculture Organization
FC Fecal Coliforms
FHI Family Health International
GDP Gross Domestic Product

GHG Greenhouse Gas
GWR Groundwater Rise
HIAs Health Impact Assessments
IEQ Indoor Environment Quality
IFRP International Fertility Research Program
IJAS Illinois Junior Academy of Science
IPBES Intergovernmental Science-Policy Platform on Biodiversity and Ecosystem Services
IPCC Intergovernmental Panel on Climate Change
IUCN International Union for Conservation of Nature
LULC Land Use and Land Cover
ODTS Organic Dust Toxic Syndrome
OECD Organization for Economic Cooperation and Development
OEQ Outdoor Environment Quality
PET Polyethylene Terephthalate
PMV Predicted Mean Vote
POP Movement Protect Our Planet Movement
POPs Persistent Organic Pollutants
PPCPs Polypropylene Copolymer
PPD Percent Population Dissatisfied
PTSD Post Traumatic Stress Disorder
REPDA Public Registry for Water Rights
ROFIJ Revolution of Food in Jamaica
SAD Seasonal Affective Disorder
SDGs Sustainable Development Goals
SDS Sand and Dust Storms
SOFI State of Food Security and Nutrition
TNCs Transnational Corporations
TSS Total Suspended Solids
UHI Urban Heat Island

UNAM National Autonomous University of Mexico
UNEP United Nations Environment Program
UNEP United Nations Environment Program
UNESCO United Nations Educational, Scientific and Cultural Organization
UNESCO United Nations Educational, Scientific and Cultural Organization
UNIAEA United Nations International Atomic Energy Agency
UNODC United Nations Office on Drugs and Crime
USMCA United States-Mexico-Canada Agreement
WHO World Health Organization
WOAH World Organization for Animal Health
YCOIL Yamaye Council of Indigenous Healers

9

Profile of Contributors

Mr. Kasike 'Kalaan' Nibonrix Kaiman, Member of Council of Ancestral Indigenous Medicine of the Americas

Kalaan Nibonrix Kaiman is Kasike (Chief) of Yamaye Gunaí Taíno Peoples (Jamaican Hummingbird Taíno People). He is also the Caribbean Region Organizer for Peace and Dignity Journeys (an inter-tribal prayer run) and holds memberships in the Council of Indigenous Traditional Healers of the Americas, Yamaye Council of Indigenous Healers (YCOIL), and the Caribbean Organization of Indigenous Peoples. Chief Kalaan is an Indigenous rights activist and promotes recognition of his people in Jamaica as well as the rights of Indigenous peoples across the West Indies and globally. He is the custodian of Yamaye/Jamaican Taíno traditions and represented YCOIL at the United Nations Committee on the Elimination of Racial Discrimination and at the Inter-American Commission on Human Rights (an entity of the Organization of American States).

Ms. Kasikeiani Chieftainess Ronalda, Member of Jamaica Council of Caribbean Organization of Caribbean Region Peace and Dignity Journeys Main Organizer

Ronalda is foremost a lover of all food and takes joy in sharing her ancestral food knowledge. She is a food and lifestyle coach, certified personal nutritional advisor, and home educator. Her interests include Indigenous home education; food security and food sovereignty; and sustainable living.

Kasikeiani Ronalda is the founder of Revolution of Food in Jamaica (R.O.F.I.J). She focuses on ancestral teachings on food as medicine. Embodying her Taino name Kaikotekina "Island Food Teacher," she dedicates her time to the development and wellbeing of present and future generations.

Dr. Kriti Akansha, Research Scientist, Mu Gamma Consultants, Gurugram, Haryana, India

Kriti Akansha is a dedicated research scientist at Mu Gamma Consultants, a leading environmental consulting firm, where she focuses on the critical issue of impact of industrial pollution on water bodies. Specializing in studying and implementing strategies for pollution mitigation, Kriti brings a wealth of expertise to her role. Holding a Ph.D. in wastewater treatment and bioremediation from the Birla Institute of Technology, Mesra, her academic background aligns closely with her professional pursuits. Kriti's contributions to the field are evident through her impressive portfolio of research publications in esteemed national and international journals. With a passion for sustainability and a commitment to finding innovative solutions to environmental challenges, Kriti plays a vital role in advancing environmental stewardship and promoting a healthier planet.

Ms. Manisha Jain, Research Scientist, Mu Gamma Consultants, Gurugram, Haryana, India

Manisha is a PhD scholar in environmental biotechnology, a postgraduate in biological science, an amateur artist, a passionate traveler with interest in adventure activities, and a keen learner of all new things. Her interest lies in finding sustainable solutions to environmental problems for developing a methodological approach and understanding on how it can be further extrapolated to other subject and geographical areas. Manisha has been involved in various research projects including the Innovation Project of the Delhi University. She was awarded a scholarship for this work. Her research experience ranges from the analysis of heavy metals to endocrine disrupting chemicals, the management of plastic and chemical waste, and the assessment of microplastics in riverine systems. In addition to publishing research articles in prestigious national and international journals and book chapters, she has contributed to a discussion paper on matching India's sanitation policies with the Sustainable Development Goals.

Mr. Kolade Victor Otokiti, Faculty of Spatial Sciences, University of Groningen, the Netherlands

Kolade Victor Otokiti is a trained research practitioner who focus on urban and regional planning and environment and development. His expertise lies in addressing complex socioeconomics and the environment issues such as climate change impacts, inequality, and urbanization impacts. Kolade has a proven track record of leading successful community development and research projects on a variety of topics including flood risk management, urban agriculture, and social exclusion. He earned his bachelors and masters degrees in urban and regional planning from the Federal University of Technology Akure and the University of Ibadan, respectively. Additionally, he holds a masters degree in environment and development from the University of Leeds in the United Kingdom.

Ms. Helen Abidemi Faturoti, Department of Urban and Regional Planning, Lagos State University, Nigeria

Helen Faturoti holds a B.Sc and M.Sc in urban and regional planning from the Federal University of Technology Akure and the University of Ibadan, respectively. She currently serves as an assistant lecturer at Lagos State University. Helen's research focuses on risk and disaster management, climate change mitigation and adaptation, and transportation management.

Ms. Vera Urtaza Reyes, Co-Founder and Executive Director, Keystone Species Alliance

Vera Urtaza Reyes, Co-Founder of Keystone Species Alliance, a non-profit dedicated to bring more awareness to the vital role that keystone species play in maintaining the structure, functioning, and stability of ecosystems. She specializes in climate change, peace, and security, human rights, sustainable development, climate change negotiations, climate change and its direct impacts on human health, conservation, rewilding, and international cooperation mechanisms.

Ms. Pooja Sharma, Director of Legal Advocacy and Policy, Keystone Species Alliance

Pooja Sharma is a rights of nature advocate with a background in civil rights litigation and corporate law for non-profits and social enterprises. Pooja transitioned into advocating for the Earth when she came to understand that our planet, our home, needed more voices to protect and restore itself to its natural state. At Keystone Species Alliance, Pooja co-leads legal advocacy efforts with the Earth Law Center for the identification and protection of keystone species in the state of Washington and co-develops the Keystone Species Fellowship Program with CoalitionWILD to support on-the-ground research on keystone species' role as natural carbon sequesters. She also specializes in nature inclusive governance and integrating system design principles into legal systems.

Ms. Claudia Fernanda Padilla Rangel, Technical Secretary, National Council of the Maquiladora and Manufacturing Export (Index Nacional)

Claudia holds a bachelors degree in international relations from the National Autonomous University of Mexico (UNAM). She entered the public sector to support the International Cooperation Management at the National Water Commission of Mexico. She provided assistance in projects related to SDG 6. She is a Special Rapporteur for the Human Right to Safe Drinking Water.

She served as head of the department of Customs Affairs with Asia, Europe, and international organizations at the Tax Administration Service where she worked with the United Nations Office on Drugs and Crime (UNODC), the United Nations Educational, Scientific and Cultural Organization (UNESCO), and the United Nations Environment Programme (UNEP) on green customs, vaquita marina rescue, and compliance with Article 24 of the United States-Mexico-Canada Agreement (USMCA).

She then became head of the department of Customs Affairs for North America and International Security at the National Customs Agency of Mexico, serving as a link with customs authorities in Canada and the United States, and liaising with the United Nations International Atomic Energy Agency (UNIAEA).

Dr. Zahida Khan, Assistant Professor, Department of Architecture, College of Architecture and Planning Ball State University, Muncie

Zahida Khan is a researcher, educator, and designer who envisions sustainable cities through human-centered design. She is a highly credentialed specialist in the areas of architectural design, high-performance building design, sustainability, and advanced building systems. She holds a Ph.D. in Architecture from the Illinois Institute of Technology in Chicago. Her doctoral study developed: (a) predictive models of human behavior in public spaces using advanced research methodologies such as computer fluid dynamics (CFD) simulation and longitudinal observation data; and (b) a novel framework for the outdoor thermal comfort integrated Human Behavior SIMulation tool called "HuBeSIM".

Dr. Piyush Khairnar, Assistant Professor, Advanced Building Technologies and Design, Huckabee College of Architecture at Texas Tech University, Lubbock-TX

Dr. Khairnar is an assistant professor of advanced building technologies and design at the Huckabee College of Architecture at Texas Tech University. He teaches at the graduate and undergraduate levels in building science, integrated systems design, and built environment performance. His research expertise is in computational simulations for performance prediction of built environments to understand user occupation and comfort. His research focuses on the nexus of natural resource extraction and anthropogenic activities and how designers can leverage advancements in digital computation to arrive at solutions that benefit society as a whole.

Ms. Fatima Khan, Researcher, Health and Psychology, University of Illinois in Chicago, Chicago-IL

Fatima Khan is an enthusiastic researcher in health and psychology. She is pursuing her undergraduate degree in psychology (pre-med track) from University of Illinois in Chicago.

Her interest includes studying students' motivation levels in education and influence of self-perception on individual's own behavior and confidence. She won a gold medal from Illinois Junior Academy of Science (IJAS) for her survey work involving sound-shape mapping (using Boba-Kiki effect) of mono and multi-lingual people residing in different part of the globe. She is a strong advocate of equity in healthcare.

Dr. Taher Mahmud, Consultant Rheumatologist, Co-Founder of London Osteoporosis Clinic; Trustee of Global Osteoporosis Foundation; Director YouOptimized

Dr Taher Mahmud is a Consultant Rheumatologist and Co-Founder of the London Osteoporosis Clinic. He has trained at prestigious institutions including the King's College Hospital, St Thomas' Hospital, and Guy's Hospital. Dr Mahmud has served as an Honorary Consultant at the Royal National Hospital for Rheumatic Diseases in Bath and as Lead for Osteoporosis and Consultant Rheumatologist at Tunbridge Wells Hospital NHS Trust. His work focuses on improving patient outcomes through holistic osteoporosis treatments and bone health advocacy. His research has explored drug side effects. He is dedicated to eradicating osteoporosis-related suffering.

10

Profile of Editors

Dr. Saroj Pachauri, Public Health Specialist, Trustee, Center for Human Progress, New Delhi, India, and Director, POP (Protect Our Planet) Movement, New York, USA

As a public health physician, Dr. Pachauri has been extensively engaged with research on family planning, maternal and child health, sexual and reproductive health and rights, HIV and AIDS, and poverty, gender and youth. In 1996, she joined as Regional Director, South and East Asia, Population Council and established its regional office in New Delhi which she managed until 2014. In 2011, she was awarded the prestigious title of Distinguished Scholar, an honor rarely bestowed.

She worked with the Ford Foundation's New Delhi Office (1983-1994) and supported child survival, women's health, sexual and reproductive health, and HIV and AIDS programs. Before that, she worked with the International Fertility Research Program (IFRP) which was later renamed Family Health International (1971-1975) and the India Fertility Research Programme (1975-

1983). She designed and monitored multi-centric clinical trials globally to assess the safety and effectiveness of fertility control technologies. During 1962-1971, as faculty of the Departments of Preventive and Social Medicine at the Lady Hardinge Medical College, New Delhi and the Institute of Medicine Sciences, Varanasi, she helped to develop this new discipline.

She has published sixteen books and contributed chapters to 21 books. She has over 100 publications in peer-reviewed journals and several articles in print media.

Dr. Ash Pachauri, Director, Center for Human Progress, New Delhi, India, and Senior Mentor, POP (Protect Our Planet) Movement, New York, USA

Dr. Ash Pachauri has a PhD in behavioral science and technology and a master's in international management. Having worked with McKinsey & Company before pursuing a career in the social development arena, Dr. Pachauri's experience in public health and sustainable development emerges from a range of initiatives. Notably, he has made significant contributions to the Bill & Melinda Gates Foundation by contributing to its public health and community agenda, the UN by focusing on youth, health, and the Sustainable Development Goals (SDGs), and the Center for Disease Control program interventions in the US by focusing on community interventions, especially for vulnerable youth. He has also been instrumental in founding and building the POP Movement and the World Sustainable Development Forum. He is a technical adviser to the World Health Organization on Self-Care Global Guidelines to support youth, communities, and global governments.

Dr. Pachauri has been a pioneer in the use of information technology for development. His innovative approaches have been key to spearheading community and youth-led self-care interventions, leading to global capacity building and adoption of self-care among youth. As a master trainer in behavior change communications and strategic leadership, Dr. Pachauri has led over 20,000 workshops, events, and global outreach to youth and communities to promote global health and climate action.

Widely published, winner of the prestigious Overseas Research Scholarship, awarded for advanced studies in the U.K., and recognized for his academic achievements, Dr. Pachauri's awards and recognitions reflect his significant contributions to the field. The United Nations has recognized Dr. Pachauri for his dedication and leadership in their flagship publication, "Portraits of Commitment," A testament to his influence in the field. In 2021, he was awarded the GlobalMindED Inclusive Leadership Award for action in Energy and Sustainability, a recognition of his commitment to inclusive and sustainable development among young people worldwide. He is an Associate Fellow of the World Academy of Art and Science, a position that underscores his academic standing. Dr. Pachauri serves on the Boards and Advisory groups of several organizations and initiatives worldwide, including the global movement on bone health, the Climate Change Coalition, and the Global Union of Scientists for Peace. He demonstrates leadership and influence in the global health and climate action community.

www.ingramcontent.com/pod-product-compliance
Lightning Source LLC
Chambersburg PA
CBHW070909270326
41927CB00011B/2503